Mnemonic

Also by THERESA KISHKAN

ESSAYS
Phantom Limb (Thistledown Press, 2007)
Red Laredo Boots (New Star Books, Transmontanus series, 1996)

FICTION
The Age of Water Lilies (Brindle & Glass, 2009)
A Man in a Distant Field (The Dundurn Group, 2004)
Inishbream (Goose Lane, trade edition, 2001)
Sisters of Grass (Goose Lane Editions, 2000)
Inishbream: a novella (Barbarian Press, 1999)

POETRY
Black Cup (Beach Holme/Press Porcepic, 1993)
Morning Glory (Reference West, 1991)
I Thought I Could See Africa (High Ground Press, 1991)

MNEMONIC *a book of trees*

Theresa Kishkan

Copyright © 2011 by Theresa Kishkan.

All rights reserved. No part of this work may be reproduced or used in any form or by any means, electronic or mechanical, including photocopying, recording, or any retrieval system, without the prior written permission of the publisher or a licence from the Canadian Copyright Licensing Agency (Access Copyright). To contact Access Copyright, visit www.accesscopyright.ca or call 1-800-893-5777.

Cover illustration: www.istockphoto.com/ekspansio.
Cover and page design by Julie Scriver.
Printed in Canada.
10 9 8 7 6 5 4 3 2 1

Library and Archives Canada Cataloguing in Publication

Kishkan, Theresa, 1955-
Mnemonic: a book of trees / Theresa Kishkan.

Includes bibliographical references.
Issued also in electronic format.
ISBN 978-0-86492-651-7

1. Kishkan, Theresa, 1955-. 2. Novelists, Canadian (English) — 20th century — Biography.
3. Authors, Canadian (English) — 20th century — Biography.
4. Trees — Social aspects. 5. Natural history. I. Title.

PS8571.I75Z467 2011   C813'.54   C2011-902889-1

Goose Lane Editions acknowledges the financial support of the Canada Council for the Arts, the government of Canada through the Canada Book Fund (CBF), and the government of New Brunswick through the Department of Wellness, Culture and Sport.

Goose Lane Editions
Suite 330, 500 Beaverbrook Court
Fredericton, New Brunswick
CANADA E3B 5X4
www.gooselane.com

For my brothers Dan, Steve, and Gordon Kishkan
"...impossible to imagine a world without them."

Contents

11 Prelude

15 *Quercus garryana*
    Fire

33 *Quercus virginiana*
    Degrees of Separation

61 *Olea europaea*
    Young Woman with Eros on her Shoulder

81 *Thuja plicata*
    Nest Boxes

105 *Platanus orientalis*
    Raven Libretto

131 *Pinus ponderosa*
    A Serious Waltz

155 *Fagus sylvatica*
    Traces

179 *Arbutus menziesii*
    Makeup Secrets of the Byzantine Madonnas

197 *Populus tremuloides*
    Cariboo Wedding

213 *Arboretum*
    A Coda

231 Acknowledgements

234 Endnotes

241 Bibliography

This is a book of memory, and memory has its own story to tell.
— Tobias Wolff, *This Boy's Life*

Prelude

> Midway on our life's journey, I found myself
> in dark woods, the right road lost. To tell
> about those woods is hard — so tangled and rough...
> — *The Inferno of Dante*, Canto 1, lines 1-3

On a forested acreage on the north end of the Sechelt Peninsula, my husband John and I built a house, raised our children, and sent them off into the world. Most mornings of my adult life, I've awoken to these woods. They are never the same though you'd have to live here to know that. For instance, in wind, the trees bend and arc wildly, some of them falling to earth but usually deep within the woods, not near the house, so visitors wouldn't know of their demise. At those times, I find myself wondering about their secret lives for it seems that they thrash and sway in joy, like Bacchantes. And stories abound of trees that live as people live, with the same sorrows and pleasures.

In winter, in snow, the evergreens are most perfectly themselves, heavily burdened. In late winter, the alder stems and buds are russet, giving way to green in spring; in fall, the alder leaves don't turn orange or yellow but dull brown, or they fall still green to the ground. Sometimes I glimpse an unexpected flutter of white down the bank below the house and eventually realize that the dogwoods are out, the bracts as large as Kleenex surrounding the small greenish-purple flowers. In fall, the leaves of the dogwoods are gorgeous, deep pink and scarlet. In high summer, no one would know that each spring the arbutus tree looks as though it's dying; this is because there are plants around it that I water over the summer and the arbutus receives too much irrigation for its liking. It does survive, though.

There are many maples in our woods, some of them mossy with age. Ten years ago, in autumn, I called a painter friend to ask, "If you were painting the bigleaf maples right now, would you use Naples yellow?" He mused for a bit, answered, "No, they're too orange," then called me back to say, "Yes." A neighbour's daughter once raced down a forest trail with a huge fallen maple leaf on her face, keeping it in place by holding her face up, laughing as she ran. A perfect disguise, I thought, so close to Halloween.

In a treatise on oratory, *On the Ideal Orator (De Oratore)*, the Roman statesman Cicero advises the training of one's memory, specific to oration and rhetoric, in a systematic and logical manner called a method of loci. By memorizing the architectural space of a particular building and by attaching elements of a speech to particular features of the building and forming an image of the two, a structural mnemonic is created. Variations of this discipline existed and persisted well beyond antiquity and traces remain in our contemporary figures of speech: "in the first place, in the second place," we say to keep our place in an argument or discourse.[i]

I had this in mind—those palaces of memory. Every time I found myself remembering, I tried to walk myself through a gracious building with nooks and window seats. Instead, I had such a clear and visceral sensation—more than a picture, more than an image—of a tree. The smell of dry grass carried memories of fire and the trees that presided over those memories were the Garry oaks of Vancouver Island where I lived as a child. Recalling a brief romance on a Greek island—I was surrounded by olive trees, their grey-green leaves shimmering in the heat of that lost time.

I also had in mind John Evelyn's *Sylva: A Discourse of Forest Trees & the Propagation of Timber in His Majesties Dominions*. Ostensibly, the book is an inventory of Britain's arboreal holdings undertaken to address the shortage of timber available for shipbuilding (and thus exploration), transportation of every sort, fuel for the manufacturing of glass, smelting tin and iron, brewing, cloth dyeing, and for domestic use; in short, the necessities of Empire. Evelyn achieves this, but the book is also a love song to trees in all their nuanced beauty. He wrote, "Here I am again to give a general notice of the peculiar excellency of the roots of most trees, for fair, beautiful, chamleted and lasting timber, applicable to many purposes; such as formerly made hafts for daggers, hangers, knives, handles for staves, tabacco-boxes, and elegant joyners-work, and even for some mathematical instruments of the larger size, to be had either in, or near the roots of many trees..."[ii]

In the first place... What do I know about the habits of trees? I'm not a botanist and barely passed high-school biology. I can name them, count them, keep lists of their occurrence in landscapes familiar and far-flung. How we once saw a Douglas fir on a ledge of rock at Island in the Sky, within Canyonlands National Park in Utah, far from its usual range. How I went to the western red cedars at the Royal Botanical Gardens, Kew, when I lived for a time in London and touched their bark for the remnant perfume on my hands. What do I know about them? I intend to find out. I intend to revisit, in memory, significant trees of my past, assisted by field guides and assorted historical texts which have informed my process of learning and seeing.

In the second place... Could I write my life by remembering the groves, imaginary or real, of my childhood, my girlhood, the painful years of young adulthood, of motherhood? Separated by time, by geography, by mythology — for who among us doesn't embroider in order to find the pattern of beauty and meaning in the plain clothing of experience? — the trees in the grove would congregate with botanical unease perhaps at first, those accustomed to the Libyan Sea touching boughs with those growing along the edges of the Salish Sea. But I think of the teaching gardens, the "physick" gardens, the utilitarian gardens of subsistence, and I hope that my memory plantation might have something to say about relationships between species without any apparent connection. Companion plantings, if you like.

My children have left home, the house echoing with their absence; yet like young trees their shadows leave us with a story told by the fire, season after season. In the shadows I see myself, dreaming my way back to the beginning. *I found myself in dark woods*...But not completely dark. Sunlight filtered through the wide boughs of maples. *To tell about those woods is hard*...I will tell of the trees, their bark and their shady leaves, each of them in its place. I have a few good guides to help me with my tale.

Quercus garryana
Fire

In early May of 2007, my husband John and I walked through the woodland below Government House in Victoria, under mature Garry oaks. Blue camas bloomed in great drifts like a dream of heaven, punctuated by wild roses and snowberry, fawn lilies, and grasses. It was sunny and warm, and the heat released a smell that transported me back decades, to my childhood.

It was 1965. My family was staying in a motel out towards Colwood. I was ten years old. My parents were searching for a rental house for us, now that we were back in Victoria after two years on the East Coast where my father's naval career had taken us. Some days they left my younger brother and me in the care of our two older brothers, who were twelve and fourteen. They hung out with some older kids—the offspring of the motel owners—and my younger brother and I found ways to amuse ourselves.

I took books into the area behind the motel, dry grassy bluffs with groves of oaks. The smell was intense—the grass, the leaves, sticky pitch from a few pines, the unexpected twist of onion as I grazed the stems of nodding onion.

I'd recline in the grass, ants tickling my bare legs, and read Nancy Drew adventures. I longed for a life so exciting — where treasure might turn up in a hollow tree or under a bridge; where villains might be thwarted by polite requests; where a girl would rise from a shaking up by an escaped convict, straighten her stocking seams, and drive away in her roadster for the next case. I was absorbing the dry heat, pollens, and odours as I read, my body resting on golden grass that flattened beneath my weight, satin to the touch. My younger brother stalked imaginary villains among the bluffs, talking to himself.

Later in the day, my parents returned to report on possible houses, urging us to gather up our swimming gear for a picnic to Beaver Lake. I was pulled reluctantly back into my family's orbit. The green camping stove was stowed in the back of the station wagon with Star, our Labrador. We ate wieners in buns spread with green relish and bright yellow mustard, and drank fruit punch from a Coleman thermos jug. Swimming in the weedy lake, swans on the opposite shore guarding their young, I marvelled at how I could remove myself so completely from my family and then return to them as though nothing significant had happened. And to anyone else, nothing had.

That summer there was a fire warning; the stretch of rainless days was making the oak groves volatile as tinder. Lightning was feared, or the careless flick of a cigarette butt. I'd lie in my bed in the motel at night under a single thin sheet, worrying that sparks would lick the dry grass into flames and rush down the bluffs to the unit where my family slept, oblivious. Sirens from the Colwood Fire Hall punctuated the quiet. I could smell the night outside, heavy with heat. The idea of fire seemed somehow inevitable as our lives changed — suspended between a house we'd left near Halifax, and the house on Harriet Road, which my parents had yet to find. I was afraid, but also thrilled with the possibility of such latent power. I knew, though I wouldn't have had words to say how, that we lived in an intensely mysterious and potent world, and the possibility — even prospect — of fire was part of that. I imagined fierce heat and crackle as flames consumed grass and brittle moss.

There is a long history of fire shaping this landscape of oak and dry grass. Northwest Coast peoples used fire to create ideal growing conditions for camas, the roots of which were a staple in their diet. The oak trees withstood the heat; undesirable species didn't. The burned areas produced healthy harvests of the beautiful blue camas flowers and their succulent bulbs, as well as acorns, for meal. (I suspect that vulnerable young oak seedlings would not have withstood the fires, however, so I have to wonder about subsequent generations of Garry oaks in these landscapes, though recent research by range ecologist Jon Keeley and chemist Gavin Flematti, among others, does indicate that compounds in smoke trigger germination in buried seeds.[1]) The Northwest Coast peoples burned after harvest, before rains, and developed techniques that used weather and terrain to their advantage.

"A perfect Eden," James Douglas wrote of the park-like nature of southern Vancouver Island,[2] a quality that Captain George Vancouver had thought natural and artful, never understanding how the effect had been achieved.[3] Before these controlled burns, lightning fires would have produced some similar results and might have inspired those early people to use fire as a way to increase camas growth. They would have observed how animals fled from fire and how this made hunting more successful. People living deeply in a place are the best observers of cause and effect, weather, fire, and harvest.

Anecdotal reports from the journals of early explorers and settlers attest to the improved berry crops—wild strawberries, currants, gooseberries, black and red caps—as well as nodding onions and the important camas bulbs. The fires also improved pasture and forage for deer.[4] We have been taught to think of the Northwest Coast peoples as hunter-gatherers, yet there is evidence of a kind of agriculture, practised with care and skill—orchards of oak yielding acorns, rich fields of root crops, and berries.

Garry oaks, or more properly *Quercus garryana*, were named by David Douglas for Hudson's Bay Company official, Nicholas Garry. There are two distinct kinds of oak woodland on Vancouver Island. One of them, the Garry oak parkland ecosystem, is deep-soiled, producing big oaks such as those of the

Broadmead meadows in Saanich, where I spent my teen years. In other areas, with shallow soils and more rock, we find the scrub oak ecosystem, a landscape closer to California's than our western temperate rainforest. The understory differs too, with snowberry, camas, fawn lilies, graminoids, and brackens populating the former and spring flowering forbs, grasses, and mosses in the latter. Fires were mostly used by the First Nations peoples to control growth in the deep-soiled areas, which is where the most potential existed for good root crops.

Sitting at a desk in the Annex of Sir James Douglas Elementary School as a child, I looked out at the familiar trees — Garry oaks on rocky bluffs below Government House — puzzling through a sentence in my reader, wondering yet again why words that looked the same sounded entirely different. I was kept in at noon one day because I argued that "food" and "good" should be pronounced to rhyme. It was not explained to my satisfaction.

When I was a child in that motel, fearful that those beautiful meadows would ignite, my reading retreat disappearing in an instant, I wonder now if somehow I was caught in a wrinkle of time when children would have lain awake in their cedar lodges, the same fear and anticipation quickening their pulses and hearts. "The fire runs along at a great pace," a newspaper article from 1849 reads, "and it is the custom here if you are caught to gallop right through it, the grass being short, the flame is so very little, and you are through in a second…"[5] Was I running with those children, our feet swift on the dry grass, flames racing ahead, and behind? Was the acorn I pocketed a descendant of one a child in 1849 might have gathered with his mother, anticipating the taste of them steamed or roasted, before the excitement of the coming fires? Or one that lingered underground, longing to be awakened by smoke?

So often a myth contains within itself a kernel of absolute truth — a codex, an epistemology: the stories of harps singing on their own, the music contained in the wood of their making; gods and druids who took their wisdom from trees; acolytes seated at the foot of a banyan, or perhaps an oak, hoping for enlightenment.

The etymology of oak is curious and revelatory. Ancient Indo-European roots for "tree" begin as oaks—*der-*, *dru-*, *doreu-*, *derwo-*—before evolving through the Gothic *tru* and Old Norse *tré* to the Anglo-Saxon *treo*. Underlying this was a belief that oaks were the most important of all trees, where sacred names had their origins. Can you hear "druid" in these roots? And the Attic Greek word for tree: δρῦς *drys*, echoes "oak." It's only a short linguistic distance from *drys* to "dryads," the feminine personification of the oak tree spirits.[6] Dryads possessed some divine gifts—specifically prophecy.

Researchers now believe that trees can hear, that receptors for sound are located in the leaves above ground. When plants synthesize the hormone gibberellic acid, it accelerates growth but also has been found to promote a listening response, the range of which is slightly louder than the human voice.[7] All those science fair experiments investigating the effect of music on pea plants had merit after all!

In thinking about this, I am reminded that oaks were venerated in Europe in pre-Christian times and were associated with various pagan divinities—the Greek god Zeus; the Celtic god Dagda; the Norse god Thor.

In ancient Greece, areas struck by lightning—frequently oaks because they were the tallest trees and poor conductors of electricity—were consecrated to Zeus. The supreme god of the Greek pantheon and god of weather (his epithets include "cloud-gatherer" and "hurler of thunderbolts"), one of his symbols is the oak tree. People heard the voice of this god, and others, in the rustling of oak leaves.

The Norse Thor was the god of thunder, associated with strength, fertility, and protection; his symbolic tree was also the oak. And Dagda was a principal Celtic god, an earth god, protector of crops. He carried a magical harp of living oak wood.

The beautiful Celtic alphabet, Ogham, is based on trees, its twenty original and five subsequent characters named for trees or shrubs;[8] and we find oak firmly within this system as *dair* or *duir*. Given the Irish love for trees and their placement within Irish mythology — who can forget the salmon of knowledge that fed on hazelnuts dropping into the River Boyne? — this alphabet is not surprising. Listen to this little Gaelic poem and its translation, both by Aonghas MacNeacail:

dh'éirich craobh	a tree arose
is dh'abair i rium	and spoke to me
bha litir na bial —	a letter in her mouth —
a samhla	her likeness
cha mhis an tùs	i'm not the beginning
ars a chiad chraobh	said the first tree
cha mhis a, chrìoch	i'm not the end
ars an litir mu dheireadh	said the last letter[9]

Years ago, I stood in the Kilmalkedar churchyard on the Dingle Peninsula in Ireland, and ran my hands over the surface of an ogham stone. I didn't know then that the stones were used as boundary stones, to mark territory. I thought it was simply a burial stone, which it may have been, as they served this purpose, too. (Most ogham inscriptions are tallies and serial groups of names.)

The stone was pierced at the top with a hole. I tried to put my hand through. Later, when I read Miranda Green's *The Celtic World*, I learned that these pierced stones were symbols of fertility, regeneration, and healing. The Christian fathers were canny enough to recognize the utility of these pagan beliefs and foundations and then to incorporate their own ikonography. So an alphabet of trees, carved in stone, standing for territory and naming, eventually enclosed in a churchyard surrounded by rowans, sycamores, and a lush *Quercus robur*, the great Irish oak — a hole in the top to ensure renewal. I walked under those trees, my hand tingling from its encounter with the stone. *A tree arose / and spoke to me, / a letter in her mouth...*

One afternoon, when I was in my very early twenties, I went with a friend to a fortune teller in a café on Yates Street in Victoria. This was in the mid-1970s, when young women dressed in gypsy skirts. Shops selling jewellery from India, crystals, and yarrow sticks for divination were clustered around lower Yates and Bastion Square.

The fortune teller was surprisingly ordinary, a stout woman in her sixties in a flowered dress, set up at a booth in the corner. I don't remember what I paid. The woman did something with cards, my palms, and the pattern of leaves in the bottom of my teacup. I wanted to know my immediate future. Was there love in the offing? Would I be happy? She gave a very general reading, not particularly vivid or inspired. But when I got up from my seat to let my friend take her turn, the woman quickly wrote something on a scrap of paper. She put a hand on my arm to get my attention and said, "I want you to call me this evening. I have something to tell you which I'd rather not tell you now, with people around."

I was intrigued. Of course I was. Was there a prince apparent in the lines of my hand? Were riches coming my way? Was it that obvious? I dialled her number that evening, too nervous to eat dinner first, and probably fortified with a glass of wine (Similkameen white in those days, purchased in jugs). I wasn't sure she'd remember that she'd asked me to call, but she immediately told me that she had sensed I was in the care of Pan. "He is watching you," she told me. "He watches out for you, and you mustn't be afraid. One day he will pass you on the street, and you'll know it's him. Remember this." And that was all she had to say. I imagined her in an overstuffed chair, shoes off and stockings eased down her heavy legs, placing the telephone receiver back on its cradle. Her knitting waiting beside the chair.

Well, this was not what I'd hoped for. And I wasn't sure I gave the conversation much credibility. But as a student of Classical literature and mythology (I had taken a number of courses in these subjects at university, enough to qualify me for a minor in Classical Studies), I knew something of Pan. God of shepherds, horned, hoofed: stories have him in pursuit of nymphs and mortals, who are turned into reeds to escape his amatory advances.

In the Homeric Hymn to Pan, the poet or rhapsode (from the Greek *rhapsodes*, "one who stitches together") reported, "Only at evening, as he returns from the chase, he sounds his note, playing sweet and low on his pipes of reed: not even she could excel him in melody—that bird who in flower-laden spring pouring forth her lament utters honey-voiced song amid the leaves."[10] And truly, who cannot claim to have heard that note, wind over the hollow grasses, just beyond sight?

Pan was to be found in remote unsettled places, rocky heights, and those who encountered him, or even sensed his presence (the rustling of oak leaves), might be filled with a feeling that became known as panic (from the Greek *panikos*: a word derived from the god Pan's name). I thought of the places I ventured alone—the highlands above Durrance Lake, the wild beaches past Sooke towards Jordan River—and how I thrived on that feeling which was not quite fear, not quite awe. Was this the god passing through the trees, just beyond sight?

I realized that Pan was also the god of the constellation associated with my astrological sign, Capricorn. When I indulged in something like belief in astrology, I'd muse on the qualities I was said to embody: steadiness, tenacity, practicality, a reluctance to forgive, to show emotion, a tendency towards convention. An interest in making money. I thought my sign was something I'd grow into because so far—I am writing here of my early twenties—I was not practical. I would spend a week's food money on books or a dress, and then eat plain boiled pasta until I could afford a carton of yogurt or a bag of apples. I wept too easily and too frequently. I did nurse grudges, akin to that lack of forgiveness, but that was hardly something to be proud of. And I wasn't sure I could claim tenacity, that sure-footed climb of the goat towards its goal.

Then one day, walking along Douglas Street by the Hudson's Bay department store, a man passing gave me a look — not amatory, but complicit somehow — and I immediately knew he was Pan. I turned to talk to him, or follow him, or somehow make contact, but he was nowhere to be seen.

When I told a friend about this encounter, I saw the doubt in her eyes and realized that this was a matter of faith for me, and I would need to learn either to defend with passion and evangelical fervour, or remain silent. Only once or twice over thirty years have I mentioned the fortune teller and my *daimon*, if he can be called that. But in a curious way I have known myself to have the guidance of Pan, an intermediary between myself and the larger world, replete with knowledge and divine power. Perhaps this is simply an acknowledgement of fate — that our days are measured and apportioned.

Socrates said that his *daimon* provided warning but didn't direct action and was more accurate than watching flights of birds or reading entrails. I have done both, traced the pattern of love in the long skeins of geese or the quick updraft of kinglets and fallen to my knees in sorrow at the spilled intestines of a young fawn still in its spots by Wallace Drive as I cycled to work at the Butchart Gardens.

"On his back he wears a spotted lynx-pelt, and he delights in high-pitched songs in a soft meadow where crocuses and sweet-smelling hyacinths bloom at random in the grass," the rhapsode sang in the Homeric Hymn to Pan.[11] And for a time, his later incarnation was carved in stone to peer out of the corners of the great cathedrals. When I see images of the Green Man, a face fringed in oak leaves, some even coming out of his mouth, his wide delighted eyes, I think of Pan and his legacy: a spirited god at large in the world, making serious work of being alive in wild places.

All those years ago, lying down among Alaska oniongrass, wildrye, long-stoloned sedge, Pacific sanicle (those footsteps of spring), the seedpods of the great blue camas, I might have been listening for the distant sound of reed pipes as my guardian spirit pursued nymphs in the hills above Colwood. The dry grass crackled and the oak leaves rustled.

It's 2007, and I live on the Sechelt Peninsula, where a friend has a Garry oak he grew from a seed. There are no others that I know of on this peninsula, and why would there be? The field guides are clear in their delineation of its range: the dry slopes and meadows on southeastern Vancouver Island; the Gulf and San Juan Islands and elsewhere in Washington State; down into Oregon, where it is the Oregon white oak. There are two small locations in the Fraser Valley where it grows—one this side of Chilliwack, and the other near Yale. I am trying to think, now, of places where I knew it in abundance.

In the late 1960s, I used to saddle my horse early on weekend mornings and ride him across the Pat Bay Highway to a gate leading up onto the old Rithet's farmland. I was in my early teens, a lonely girl in search of lonely places. Someone had told me that it was fine to ride there, but that the gate had to be kept closed, as there were cattle grazing in the area. I don't really remember the cattle, but I occasionally saw deer in the tall grass. There were many oaks growing on the slopes. In the spring, there were expanses of blue camas, yellow buttercups, and odd brown speckled flowers that I now know were chocolate lilies.

I loved the open beauty of those meadows, where pheasants roamed and flew up, sharp-winged as we approached. The meadows smelled intensely dry, fragrant as hay, though not dusty. I'd let my horse canter up the long slopes and loved the way sunlight filtered through the trees.

I imagined those fields unchanged since the dawn of time. Yet now I know that the area was once Broadmead Farm, where Robert Rithet bred and raised his prize thoroughbreds. There were barns, paddocks, grooms for the sleek horses—even a racetrack farther up, near where the Royal Oak Burial Park is now.

Grazing changes a place. New forbs and grasses come in hay, some of them invasive. Thistles and mustards were introduced, along with persistent yellow broom and tenacious Himalayan blackberries — so succulent, yet undesirable because of their ability to seed themselves prolifically in the droppings of those who feast upon them.

When I rode my horse over Broadmead in the late 1960s and early 1970s, the land was still owned by the Guinness family in Ireland. Oak meadows stretched back almost to the Blenkinsop Valley, the hills eventually rising to Mount Douglas. A low area, fringed with the native willows and hardhack, was used for growing potatoes. Gulls and other water birds flocked in multitudes during winter, when the land tried to return to its boggy origins: black soil immersed in cold water.

The Guinness family donated the bog — forty-two hectares — to the District of Saanich in 1994; it has been restored to something like its earlier state. A trail surrounds the bog. I've walked there in recent years, listening to red-winged blackbirds and various warblers, heartened to see tiny oak seedlings planted here and there, protected with wire cages. But the meadows are gone. They have been developed into big houses and a few modern churches. For me, there are ghosts. If I look quickly, I see the sere grass and the big oaks, their gnarled branches bare in winter. If I close my eyes, I hear the sound of hooves as a girl races her horse up the slopes, pheasants rising in panic.

And I long for the sight of the western bluebirds and marble butterflies that were once a part of those meadows. The golden paintbrush and the Lewis's woodpeckers, now extirpated. There were snakes and lizards on the dry rocks in those days, and while some may still remain in isolated pockets, there are no longer the healthy populations that ensure survival of a species. Such populations depend on an elegant, balanced symmetry.

I could never pass the big oaks of the old British Columbia Protestant Orphans' Home on the corner of Hillside Avenue and Cook Street without wondering what it must be like to be abandoned by parents, by death or poverty or illness.

My home life wasn't perfect—I was an avid reader, and imagined perfection to be something like those stories where brothers and sisters were kind to one another, fathers never shouted, mothers never called a girl ungrateful, and harmony reigned in the house and garden.

I had two parents and three brothers. Things felt normal in most respects. Some days, I would imagine myself to be an orphan, and could work up a considerable amount of self-pity, creating a story in which I was given to cruel guardians and worked to near-death in the kitchen. If I were asked to do something beyond my usual chores at home, I'd lean on my rake or hoe as I took a break from raking the grass or weeding the carrots. Tears would course down my cheeks as I swept the driveway or walked over to the grocery store for a pound of margarine or several tins of tuna for the casserole my mother was planning for dinner. I managed to produce ample indignation at the ways in which an orphan could be exploited.

The Orphans' Home was built in 1892 with funds donated at the bequest of John George Taylor. It opened officially in 1893. Until then, orphans had been lodged first in the private homes of Mary Cridge and other compassionate citizens of Victoria, and then in a cottage bought for that purpose on Rae Street in 1873.

Curious, a few years ago I researched the Orphans' Home, and found an application form for the admission of a four-year-old boy, signed by his father, attesting that he was healthy enough to enter the Home. What was not said was why the father was placing him there in the first place. The father's place of residence was Nanaimo, so perhaps he was a coal miner. Maybe the child's mother had died and the father had no interest in raising him alone.

Other letters reveal family violence, abandonment, and destitution. Concern is expressed for a nine-year-old girl (a "half-caste") who seemed destined for the streets if a place could not be found for her at the Home. A mother wrote to say she needed help with her four children, their father having left the country, and in any case, she had discovered he had another wife and family elsewhere.[12]

From the holdings of the British Columbia Archives, I also read a poignant exchange, undated, between Flora Sinclair and Mary Cridge, the wife of Bishop Edward Cridge:

Dear Mrs. Cridge,
    Would you kindly let me go to work at a situation for a change because I have been here ten years on the 30th of December and I wish very much to go away.
    Yours sincerely,
    Flora Sinclair

And the reply:

Dear Flora,
    I read your request to the Ladies of the Committee today. We all agree that if you wish to leave the Home to work outside you may go when a proper situation is found for you. Would you like to have the charge of children or do you prefer housework? Remember dear child our wish is that you may be happy and useful and you will if you go where God directs.
    Your [sincere?] friend,
    Mary Cridge

These letters tell me a few things. Flora must have come to the Orphans' Home as a result of a death, abandonment, or family hardship in the darkest days of the year. How sad to find oneself settling into an orphanage just after Christmas, as the year was about to turn. As for the year, I have no idea, but Mary Cridge died in 1903, and the whole enterprise of housing orphans began in 1873, so I'm assuming this was late in the 1800s, after Flora had endured a ten-year residence in the Home. She could not have been particularly happy there: "I wish very much to go away." She does not thank Mrs. Cridge for watching out for her welfare or providing a roof over her head (the meals listed in the Orphans' Home archive are heavy on boiled meat and soup, relentlessly mundane).

What does Mrs. Cridge offer? Childcare or housework. What if Flora had been a talented artist or had the gift of perfect pitch or wanted to become a nurse? No, she must go where God directs, or in this case, where His proxy, Mary Cridge, decrees. Flora's handwriting is neat and tidy, each word formed as though in response to her need to state her wish as clearly as possible. Mary Cridge's reply is scrawled, the writing of a busy, elderly woman, perhaps taken aback at the forward nature of Flora's request: "Would you kindly let me go…"

My Girl Guide pack was taken to the Orphans' Home in the mid-1960s. I don't remember the point of our visit — a kind of service, I suppose, and no doubt badge-oriented, as most of our activities were, from learning to cook a hamburger on a little candle flame, to knots, making bivouacs, and other camping skills of Boer War vintage — but I do remember that it was awkward. Our Guide pack was comprised of girls from a neighbourhood near Tillicum Elementary School, mostly from two-parent households with assorted siblings, and membership on various school teams. We were old enough to roll up the waistbands on the skirts of our uniforms slightly, to show a bit more leg. Some of us wore eye makeup and Cover Girl liquid foundation (I couldn't include myself among these femmes fatales; my mother would never have permitted it, and I had yet to discover the delicious and risky habit of defiance). I was still impressionable enough to want to do good deeds in my community (and earn a badge to add to my sleeve). The orphans were reluctant to mingle with us and do the craft we had brought to demonstrate — would it have been braiding key holders with plastic cord or knitting a dishcloth out of Phentex? The songs we'd practised to sing as rounds were not the success we'd expected them to be, either. I remember asking one girl about her family. Her sullen reply was that it was none of my business.

The Home was — and continues to be (for it still exists, though no longer as an orphanage but as the Cridge Centre for the Family, an ironic twist) — a big red brick Edwardian pile, standing on a hill with rocky bluffs on the Hillside exposure. The oaks are magnificent; tall and proud, survivors of ancient fires. I always wanted to walk up among them, but a chain-link fence surrounded the complex, adding to its forbidding aspect. In the archival material, provided at a Web site devoted to the history of the Orphans' Home, there are strict rules about visiting times to the Home, and perhaps our Guide leaders were given a small window for our projected visit.

Ten years later, in winter, I'd sometimes pass those bluffs with their oaks on my way across the city and see the pattern of gnarled branches reaching up for the grey sky. I thought of quiet elk, antlered and alert, hoping to pass safely through the darkest months. I thought of the slow incremental growth of bark and the strands of lichen. I wondered if Flora ever wandered among them on her own, in her tidy dress and apron, or if small boys played hide-and-seek with true abandon on those grounds.

As a young adult, I had an apartment on Fort Street. Shadows of oaks patterned the walls of my small sitting room. It was comforting, as though that herd of elk had settled nearby to wait out winter, listening for danger, their bodies warm and familiar. I was waiting out winter myself, hoping to return to a life I'd begun with a lobster fisherman in the west of Ireland. That changed when I met a poet in Victoria and fell in love with him. Thinking and weeping through this in my apartment, I'd stand by the window and look out at the twilit street, the oaks with their heavy limbs beginning to unfurl chartreuse leaves, their clusters and catkins of flowers, female and male, almost invisible.

I'd walk out in the darkness to sort out my feelings. Up through Rockland and down into Fairfield, where the sea tossed fretfully beyond Dallas Road. These were the city streets I was familiar with, lined with chestnuts and flowering plum and cherry, with clusters of oaks on rocky outcrops. The oaks were the wild trees,[13] even in their placement on the wide residential avenues where they'd been built around or accommodated for. In some cases they'd been left on purpose, as in the meadow behind Government House where very old ones, with massive crowns, stood in timeless repose with their distant view of Juan de Fuca Strait.

When I was a teenager, in the years when I rode through Broadmead, my parents lived at Royal Oak. There were many oaks in that neighbourhood, from the contorted beauties gathered around the Maltwood Art Museum, to those on Beaver Lake Road and other back roads off West Saanich Road. My terrain was wide in those days. On Saturday mornings, I'd tack up my horse and ride out onto Saanich Peninsula, taking the quieter roads and trails as far as Island View Beach where I'd let my horse gallop along the sand. In early summer, I'd remove his saddle and ride him into the water where he liked to plunge and snort. Depending on the tide, sometimes he'd even swim with me clinging to his back. I'd walk him on the beach to dry him off before saddling him again to ride back along Mount Newton Cross Road and through a network of trails, past Bear Hill and Elk Lake, and finally back to the rented field and barn overlooking Pat Bay Highway.

I have a map, prepared by HR GiSolutions Inc. in 1997: *Historical Garry Oak Ecosystems of Greater Victoria and Saanich Peninsula.* Green areas indicate 1900 ecosystems that cover nearly half the map's area. Red dots indicate 1997 ecosystems—a tiny smattering, decorative on the map, but shocking when one reads this as contrast, as loss. I wondered how the earlier range was determined, and asked Ted Lea, the field ecologist who helped prepare the map. He generously responded by email:

> Much of the map is derived from original lands surveys that, at least for most of the CRD area, show what was coniferous, what was broad-leaved or prairie (with oak regenerating after aboriginal burning stopped) and what was wetland.
> Otherwise, I used old photographs, paintings, soils maps and drove every street in the area to see where oaks presently exist and might have existed in the 1850s and 1860s. There are other sources

such as First Nations people and land survey journals that would have helped, but would have been too time consuming; however, this would have been fascinating to research.[14]

I was curious, too, about palynology, the sedimentary record of plant pollens and macrofossils, and what it might tell me about the vegetation history of the area where I grew up and learned to love the oaks.

A fascinating essay, "An Ecological History of Old Prairie Areas in Southwest Washington" by Estella B. Leopold and Robert Boyd, from *Indians, Fire and the Land,* offers a deep portrait of post-glacial vegetation in a particular landscape, taking into account historical and anecdotal material regarding land use. Camas and various *umbelliferae* pollens peak in association with local fires. Douglas fir pollens drop to reflect logging and deforestation due to European settlement. Narratives of shifting relationships can be read by reconstructing plant communities, estimating climatic and environmental conditions. Charcoal shows up to tell of fires. People shape their landscapes to accommodate what they require — and not just the First Nations people who tended their camas crops and their nodding onions, and who knew that ash improved the growth of wild tobacco.

People arriving later from elsewhere often brought mementoes or improvements, uncertain that the place itself would be adequate. Think of the English sparrow, the European starling, Himalayan blackberry, gorse and Scotch broom (all the widespread and invasive offspring of three seeds that germinated in Sooke in 1850),[15] the beautiful, soft-eyed fallow deer seen on the Gulf Islands and the Saanich Peninsula, to which they swam from James Island at low tide. I saw fallow deer on Island View Beach, making their delicate way through the flotsam and jetsam at the tide line.

The world contains such archives — plants, birds, foreign and native, their bones and stems and pollens anchored in the sedimentary layer. The one I read in memory is almost lost; houses crouched over the vanished grasses where the oaks once listened for fire, the promise of renewal in its dense heat.

Those grey trees on the long walk home from school, down Haliburton

Road and along Elk Lake Drive to Royal Oak. How they mirrored the angst of a girl at odds with the social world—those radiant groups so tightly guarded that no one new stood a chance of belonging; the other group that would have me and took me on Friday night hunts for beer and hash, followed by pizzas in the small hours, and then headaches the next morning as I cleaned my horse's stall and wondered about dying.

The trees, presenting gnarled fists to the sky! The darkness of their bark, their sombre postures! I wanted too badly to know the world beyond the present, and I don't think I meant heaven. I wanted to know the great lively spirit that caused the tides to turn in their season, the passing of geese in the high flyways, muttering amongst themselves as they flew to warmer sloughs and lakes, the brief luminescence of fawn lilies by the trail down to Quicks Bottom, their petals turning up like Turks caps after pollination. It made me cry, this beauty, and I had no way yet to express what I felt in the face of it. I'd go out at night to visit my horse in his wide field, his black body mysterious in the dark and his white stockings glowing if there was moonlight, and I'd cry against his warm flank.

The young never know that vast and splendid lifetimes await them. Travel, lovers, children, sorrow, loss, the beauty of mornings seen from hotel windows while a cup is cradled, the scent of jasmine filling the room from an open window. Or a young woman walking the dark streets, having met a poet with whom she was almost certain she'd spend the rest of her life, trying to see stars through the tangled branches of the great oaks, their roots deep in sediments of pollen and ash. A new moon waited.

Quercus virginiana
Degrees of Separation

> To sum up the outward madness of nations, this is the land to which we drive out our neighbours and dig up and steal their turf to add to our own, so that he who has marked his acres most widely and driven off his neighbours may rejoice in possessing an infinitesimal part of the earth.
>
> — Pliny the Elder, *Natural History*

I realized, as soon as we drove down Dallas Road, in the winter of 2009, that this couldn't be the house. The numbers weren't right. I thought—I *hoped*—that the house I remembered from a Brownie field trip in 1962 might have belonged to the Newcombe family. A very old man met the Brownies at the door and led us through a dark wood-panelled hall into a large room where he showed us cases of spiders and butterflies. I remember Native masks on the walls, and a small totem pole in one corner of the room. There were rattles, bearing fierce faces of ravens and loons, which we were allowed to shake. The old man had been a missionary. We were also told he'd known Emily Carr, a name that stayed with me, although at the time I had no idea who she was. In those years, her work hadn't achieved the iconic status it properly enjoys now.

I have a clumsy theory about degrees of separation working vertically as well as horizontally. That we can trace our relationships down the rich road to the past, like an archaeologist examining the layers of Troy, the same way we can connect ourselves to others through the present. I'm not sure that six degrees will always take us somewhere significant, but in many cases it can.

Charles Newcombe was a naturalist and ethnologist (though his primary training was in medicine; he was also British Columbia's first psychiatrist). In the late nineteenth and early twentieth centuries, he acquired artefacts from Northwest Coast cultures for many museums in North America as well as England and Germany. His son William was also a collector as well as a friend of the painter Emily Carr. Charles had died in 1926, but I thought William had lived into the 1960s. I imagined it would be somehow significant to know I had been shown the collections of two men who had been so instrumental in gathering the various histories of British Columbia: its botanical record, the material culture of its First Nations, and the artistic legacy of Emily Carr. So, on a trip to Victoria, my husband and I drove over to Dallas Road, at the point where Eberts Street joined it. I had a copy of *Exploring Victoria's Architecture*, published in 1996, that featured a photograph of the Newcombe house and provided its address. We peered at house numbers. I was trying to wrestle my vague memories into something resembling fact.

But it became clear that the Newcombe house was farther west, near Ogden Point, judging by the number given in the book: 1381. As far as I could remember, the house we'd walked to with our Owls from the church basement at Five Corners, where my Brownie pack met weekly, was between Moss and Linden Streets, near Clover Point. We wouldn't have walked to Ogden Point, a group of six-year-olds on an after-school outing.

In 1962, Victoria was a city of retired military men and ladies in white gloves. W&J Wilson Fine Clothiers sold them tweed jackets and Shetland sweaters, and Murchie's carried the tea they liked, measured from great tins with a metal scoop. The Bengal Room at the Empress Hotel knew how to mix their drinks — Singapore slings, Pimm's cup, and Tanqueray G&Ts. On windy days, couples dressed as though for church walked the seafront along Dallas Road with dogs straining at leashes. My father called it their "constitutional," a word that puzzled me in this context because I thought it had to do with good government.

In those years, a child could ramble freely around Fairfield neighbourhood and the waterfront along the Juan de Fuca Strait. I had a small, blue two-wheeler, given me for Christmas in 1960, and I remember riding as far as Beacon Hill Park. A trail known as Lover's Lane was dense with snowberries, and once I accidentally knocked down a wasp nest, crying out as its angry residents stung my legs again and again. My pedal pushers had to be cut off and for days my calves were too swollen to move them much. Rather than fear, I remember extreme impatience that I had to pass the days inside when the whole world went on without me.

Near our home on Eberts Street was Clover Point, a peninsula jutting out into the Juan de Fuca Strait. A road perambulates around its circumference and in a grassy area in the middle we once stood with hundreds of others, hundreds more on the ocean side of the road, watching the Queen being driven by slowly, her gloved hand waving, first to one side, then the other. In my family, it was insisted that she waved specifically at my younger brother, but there is no way of proving this. In those years, he was skinny and all nose, and I don't imagine he stood out of the crowd to inspire special notice. Certainly we didn't present flowers. I remember standing in that grassy place with my family, feeling strange in my Sunday dress and coat.

We often walked down to Clover Point on a Saturday morning to collect bark in grocery bags for the wood-burning kitchen stove in our house. Our mother organized these outings, in any weather, insisting that we wear our oldest clothing—patched dungarees and faded kangaroo jackets. We'd walk the beaches, choosing pieces that would fit in our stove, and we'd trudge home with the heavy bags of damp bark. It was piled on the back porch to dry, within easy reach of the kitchen.

> Charles Newcombe, doctor, natural historian and anthropologist, was commissioned by Kew to collect Aboriginal artefacts from British Colombia. The objects, including fish nets and hooks, ropes, garments, baskets, woodworking tools and gambling sticks, reflect the daily life and industries of the Aboriginal peoples and hint at their extensive knowledge of the natural environment and its resources.
> —From the Royal Botanic Gardens, Kew, Web site

I'd seen the photograph of the Newcombe house in *Exploring Victoria's Architecture*, and I wanted to try to recover that moment in my childhood when I might have been a degree closer to the history I have always gravitated towards. John and I drove west on Dallas Road, past Cook Street, past Beacon Hill Park and Finlayson Point, past Mile Zero of the Trans-Canada Highway (where Douglas Street meets Dallas Road). And then I recognized the house from its photograph.

We parked our car and stood on the other side of the road. I was gesticulating with my hands in my usual excitable fashion and my husband said, "Someone is waving to you." And sure enough, two men in the glassed-in verandah were smiling and waving. An elderly woman waiting at the bus stop beside where we stood said, "It's a halfway house, you know."

I didn't know. I asked her if she knew it had been built for Dr. Charles F. Newcombe, a man indirectly responsible for the Royal British Columbia Museum. "No," she said, "how interesting. The neighbourhood has changed, of course."

As had the house. That verandah hadn't always been glassed-in, though the balusters, generous eaves, and angled bay windows hadn't been altered. I was reminded of eighteenth-century houses, Italianate in design, that I'd seen in leafy boroughs of London. A huge tree, with a crown of at least fifteen metres, spread over the front yard of the Newcombe house. It was February the first time we parked opposite the house, and though a few snowdrops and crocuses were in bloom in some of the protected gardens, it was still winter and not even local forsythias were in bud. So I was surprised to see that this tree was fully in leaf. But it wasn't a conifer or any broadleaf evergreen that I recognized. It had vaguely elliptical leaves, glossy on top and slightly downy on the undersides. It was very lovely. Was it some kind of eucalyptus? It had no odour. I pinched off a small branch so I could try to identify it at a later point, placing it in my copy of *Exploring Victoria's Architecture*, a book that had alerted me to the possibility that this might be the house of my Brownie memory by providing the wrong address: 1381 instead of 138. (Small typo + big hope = inevitable disappointment.)

I kept the house in mind and the sprig of leaves tucked into *Exploring Victoria's Architecture*. Periodically I'd look through my tree books and try to figure out what kind of tree shaded the south-facing verandah with its Georgian Revival-influenced windows and the balustraded roofline of the house built in 1907 for Dr. Charles Newcombe and his family. A tree reaching deep to anchor itself to the earth, reaching for water and nutrients, just as I tried to anchor myself in the rich loam of history and narrative. I wondered what it might mean that the tree was not a native species but something chosen for its exotic qualities, its reminders of foreign travel, its ability to conjure a landscape with resonances perhaps now mute to our century.

It never occurred to me in childhood to wonder how Clover Point got its name. The grass in the area between the ring of paved road was short and wiry, shorn by flocks of geese that patrolled the waterfront. Their droppings were everywhere. Masses of broom on the cliffs above the water hummed with bees in spring. But clover was not much in evidence. There's a story here, I mused, and went in search of it. One of its voices was that of James Douglas, who arrived at this Clover Point in 1842:

> Both Kinds [of soil], however, produce Abundance of Grass, and several varieties of Red Clover grow on the rich moist Bottoms.... In Two Places particularly, we saw several Acres of Clover growing with a Luxuriance and Compactness more resembling the close Sward of a well-managed Lea than the Produce of an uncultivated Waste.[1]

Botanist Dr. T. Christopher Brayshaw, Curator Emeritus of the Royal British Columbia Museum, suggested that the most likely clover species was *Trifolium wormskjoldii*, or springbank clover.[2] This plant's long, fleshy rhizomes served as an important food source for many Northwest Coast peoples, including the Lekwungen, who managed extensive areas of rich growth along what is now Dallas Road. Notions of gardens and husbandry vary from culture to culture, and it seems that colonists arriving on Vancouver Island didn't recognize these systems of maintenance and use (and of course it didn't occur to them that ownership of these lands might in fact be an issue). Once Fort Victoria was established in 1843, there was industrious clearing and planting of crops the new arrivals couldn't imagine a civilization doing without: carrots, turnips, potatoes, oats. What had been admired about the area—the park-like Garry oak meadows, the tall grasses, ferns, blue camas, "the several Acres of Clover" —was replaced with farms, a rifle range, hotels, and the Ross Bay Cemetery, on the site of Isabella Ross's farm.

At the turn of the twentieth century, a young man biked down Cook Street, "a rutted, muddy roadway with a large pear tree in the centre..."[3] (I close my eyes and try to see this. I wonder where the pear tree would have been. I'd like to think of it on the section of Cook Street where the small shops and fish and chip enterprises cluster now near May Street and Faithful.) That same young man remembered Sikh cremations on the beach at Clover Point, two-metre log segments criss-crossed to form the funeral pyre, with the body placed gently on top.[4]

In the early 1960s, one would have been hard-pressed to find an area of *Trifolium wormskjoldii* at Clover Point, though I remember finding condoms on the beach, washed in on the waves that also carried raw sewage from the outfall. I didn't know what a condom was, and thought it was a sturdy balloon until my mother smacked me for putting it to my mouth. She refused to say why. Many years later I realized my error and cringed at the thought. But of course the salt water would have long washed away the little sac's former contents. My oldest brother remembers catching tiny fish in tide pools near Clover Point and bringing them home, laughing as he recalled our mother's horror at the sight of him carrying six or eight fish into the house, each swimming in its own condom from the beach. We'd been told they were "dirty filthy things" and when my middle brother reported a sighting of one in our own family's toilet, our mother insisted he was wrong.

Growing up, I remember the elderly couples at work in their gardens, tending neat English borders of perennials, trees pruned within an inch of their lives, watched by a cocker spaniel or Jack Russell. These couples were kind to children whose baseballs landed in their backyards. And there were also the widows. Invited into their houses, I was filled with the sense that time had stopped. Such still quiet—I came from a home loud with four children. I realize, now, that some of these women had lost their husbands in the Great War. Photographs of smiling men in uniform, many of them on horseback, filled mantelpieces and the tops of china cabinets. My mother knew them all, it seemed, and offered my brothers and me for chores, errands, and sometimes just for company. One of the widows (though not a war widow), a Polish woman who spoke almost no English, lived on May Street in a tall ramshackle house. My mother befriended Mrs. Ciechanowski, maybe because my father could speak a few words of Polish (learned from his older half-sisters whose father was Polish; his mother's second husband, his father, had come from Bukovina and spoke Ukrainian).

We were timid about entering Mrs. Ciechanowski's house. It smelled strong. She cooked food my father raved over, dishes his own mother must have cooked for her first husband and then continued making for her second family—cabbage rolls, soups with peppery sausage and potatoes, bowls of pickled beets that dripped and stained like blood. This contrasted so sharply with our macaroni and cheese, our roast beef on Sundays. Once, my father drove us all to a cemetery in Colwood, where Mrs. Ciechanowksi's husband was buried, and she arranged my brothers and me around his grave and led us in a song that I only realized halfway through was "Happy Birthday." Afterwards there was cake and juice served on a blanket she spread over the grass between graves.

This was the golden time of my childhood. When I walked with my Brownie pack to Dallas Road and into the missionary's house. When we no doubt received a badge to indicate we'd been courteous to an old man who showed us cases of hairy spiders, artefacts of dark wood and bone, masks featuring strange birds hung with hair that looked suspiciously human, and who hoped we might understand something of the dimensions of a life spent among the heathens, urging them to accept God into their hearts.

Two years later, when my family moved to Halifax, we stayed with my grandmother and aunt in their big house on Walnut Street while my parents looked for a place of our own. I remember telling them that I wanted to be a missionary when I grew up. I don't think I was inspired by the man on Dallas Road, but was trying to rehabilitate my image which I'd felt had been tarnished by my excitement at being allowed to choose a Barbie doll once we'd actually arrived in Halifax after a long camping trip across Canada. Barbies were new on the market, and I wanted one badly. I suppose I must have expressed this to my grandmother and aunt. And when I was finally taken by my parents to Eaton's to buy one — an ash-blond model with a bubble cut — I was thrilled beyond imagining.

But when I showed off my prize to my grandmother, her disapproval was immediate: "*This* is what you were so excited about? I expected a baby doll." I suppose the doll's bust and worldly wardrobe were not considered appropriate for a girl of eight. Or those long, long legs, and feet designed for high-heeled shoes, some of which were trimmed with feathers and glitter. The house on Walnut Street was not one that had known contemporary children.

"I want to be a missionary when I grow up," I told them, desperate for their approval, their affection. Forgetting the doll for a moment, they smiled at me. My mother was visibly relieved.

But before Halifax, before we moved from Fairfield, when we were given the freedom to roam the neighbourhood on bike, or on foot, sometimes we visited the Museum,[5] then in the basement of the British Columbia Legislative Buildings. There were stuffed animals with staring glass eyes, cases of textiles, and dust. My oldest brother told me recently that he would ride his bicycle there with a couple of friends, and Dr. Clifford Carl, then the director of the Museum, would come out from his office and ask, "What can I show you boys today?" Out would come drawers of butterflies, each fixed with a pin to a display board, or a collection of rocks, or some animal fresh from the taxidermist.

More interesting to me was nearby Thunderbird Park, where a child could watch an old man at work on a totem pole while fragrant curls and chips of cedar drifted from his tools. Sometimes he talked to us, his words slow and almost foreign to our ears. I imagined he'd always been there, a relic from the days when all of Victoria was Lekwungen territory; imagined that when the city had been developed, he'd somehow been allowed to keep his lodge. When the wind was right, he could smell the sea, and gulls whirled in the blue sky above. We knew so little about the people who had lived on Vancouver Island before Hudson's Bay Company Chief Factor James Douglas stepped ashore in 1842, followed by his men, and the wheels of change began to turn.

What they brought
Japanese flowering cherries, dawn redwoods, monkey-puzzle trees, blue Atlas cedars, English walnuts, Portuguese laurels, *Pinus nigra* from Austria, Aleppo pines from the Mediterranean, Korean cedars, true and weeping cypresses, *Quercus robur* (especially the Coronation oaks in churchyards and schoolyards, in public and private gardens), bigleaf lindens, Camperdown elms, catalpas and empresses, medlars, Tasmanian eucalyptus.

Brides, a hundred of them, sailing from England on the *Tynemouth*, the *Alpha*, the *Robert Lowe,* and the *Marcella*. Some of them becoming the wives of prominent men — Emma Lazenby who married David Spencer, whose store on Government Street was where we always went for school clothes; Margaret Faussett who married John Jessop, both of them teachers, John rising to the position of Superintendent of Education for British Columbia.

Cloth, seeds, racehorses, fine English china, hats and woollens, cricket bats, boatloads of assorted dry goods for provisioning shops, surveying equipment, the language of litigation, roses and sabres and knighthoods.

Grey squirrels to displace their smaller cousins. Starlings. House sparrows. *Rattus norvegicus*.

Names. Victoria herself. Albert (for Albert Head), Saxe Point, James Bay and Langford and Colwood (nostalgic for home). The beautiful Selkirk Water and Finlayson Arm. Ross Bay and McNeill Bay, the long view from Mount Douglas and Mount Tolmie, their slopes blue with camas in spring, fringed with erythroniums. Streets named for architects and mayors and wives of merchants. Nods to the Native people themselves but in an orthography we can all pronounce.

Stories at every turn, to be remembered on maps: Glimpse Reef, for the shipwrecked barque off Clover Point in 1860; farms nestled within the names of bustling communities: Craigflower, Strawberry Vale, Broadmead, the ghosts of its horses galloping over golden grass where the smoke of burned Garry oaks hangs in the air like spiderweb.

Expectations of commerce, cotillions, academies of learning where Greek and Latin texts jostle the butterfly nets at the ready for species now extirpated or worse, extinct. *Euchloe ausonides*, the Large Marble, *Euphydryas editha taylori*, Edith's Checkerspot, and *Callophrys mossii mossii*, the nominate subspecies Moss's Elfin.

The shoemaker, John Fannin, with his menagerie of stuffed game animals, hungering for a museum.

A version of Wawadit'la
The old man we watched carving at Thunderbird Park, a man with a soft voice and younger helpers, whose tools left curls of cedar behind — spicy *Thuja plicata*, or western red cedar, and the skunkier pungent *Chamaecyparis nootkatensis*, or yellow cedar — might have been Mungo Martin. The time was right. Mungo Martin and his son-in-law Henry Hunt worked on the Welcome sign posts between 1960 and 1962, the same years I was riding my blue bike through Fairfield and into Beacon Hill Park. The ceremonial big house at Thunderbird Park is a smaller version of Martin's ancestral property at Tsax̱is, or Fort Rupert, a house called Wawadit'la, which belonged to the hereditary chief Nak̲ap'ank̲am. As the inheritor of that title, it was Mungo Martin's right to build the house and to display its associated carvings. The house was dedicated in a ceremonial way, with potlatches and dancing, in December 1953.

While the city buzzed and hummed with the electric wealth of the postwar years, a man, not Lekwungen at all but from Kwakwaka'wakw[6] territory, patiently carved sea lions and grizzly bears in a smaller version of his home on the northern end of Vancouver Island. And in any place, Tsax̱is or Victoria, in any era, seventeenth century or mid-twentieth, there would have been children drawn to the smooth movements of his hands, fascinated but not surprised to see the strange animals emerge from the wood.

This was a man who'd seen too much of his material culture disappear to anthropologists, museum collectors, and missionaries who were keen to remove the vestiges of a spiritual life they couldn't begin to understand, and yet were not averse to making money from them afterwards. Teams of authorities seized ceremonial items used during potlatches or asked for their surrender as grounds for suspending sentences. Mungo Martin was given the task of recreating many of the poles and canoes and masks that had deteriorated beyond repair. (I think of the young men who carved with him: his son-in-law, his grandsons, others, all of whom learned by watching his hands, felt the weight over theirs as they used adzes, knives, and chisels, learned how to work with an imperfection, how to judge depth and grain.) He was given a small corner of the colonial enterprise on Victoria's Inner Harbour to build the scaled-down version of

Wawadit'la and to bring together elements of the coastal cultures so assiduously removed from the city in its earlier days. (In Victoria itself, the Lekwungen village site at Songhees was relocated, the waterways and wetlands once used for travel filled in or directed underground through culverts, the language discouraged, the spiritual practices vilified.)

Those degrees of separation... When I was a young teenager, and my family lived at Royal Oak, I often rode my Anglo-Arab gelding down our street to a series of trails on Colquitz Creek. There was a house tucked back off the road, a sort of colonial-style bungalow; two more of a similar design occupied the front part of the lot nearer the street. At the end of the driveway leading back to that house, I'd sometimes see an elderly man out for a walk with his daughter and wife. They were the Footners. I remember the man telling me I had a fine animal.

Once I was helping my mother go from door to door in our neighbourhood on behalf of the March of Dimes, and the Footners invited us in while they found a contribution for our collection. It was an interesting house with photographs and chintz curtains, I remember, and sat in its privacy with old roses and apple trees around it. Many years later, while engaged in research for a novel I'd set in the orchard settlement of Walhachin, I was astonished to discover that elderly man, Bertram Chase Footner, had been responsible for designing and building the houses of that settlement. I'd been there for research and had been taken by the remnants of the old community, a few of the bungalows with their steep-pitched roofs, high ceilings, and wraparound porches attesting to careful attention to climate. His own house there — for he'd lived for a time at Walhachin; his daughter Mollie had been born there (the middle-aged daughter I'd known a little in my teenage years) — was built of river-stone and still sits quietly elegant above the Thompson River.

If I'd known then what I know now, I'd have asked Mr. Footner questions. I'd have asked about his life after the Boer War when he built bridges in the

Sudan, a time and place so far away from our semi-rural street in Royal Oak, named for its groves of Garry oak. But he died in 1972 before I knew I would go on to write books, that I would be passionately interested in the history of the province I'd been born and raised in and took for granted until my own middle years.

I've tried, not hard enough perhaps, to find out if he had built the house he lived in at Royal Oak and perhaps the other two that resembled it on what might have been a larger lot that he'd subdivided. One archivist I spoke to insisted that the street hadn't existed before the 1950s. Yet there was an ancient farmstead across the road when we first moved there, with an equally ancient apple orchard and cider press. The owner, Bill Mahon, told my parents it was the oldest house in Saanich. It was torn down in the late 1970s for a subdivision. Well, maybe not under the current street name, I suggested to the archivist, but the road itself certainly existed well back into the century and maybe before. He wasn't convinced.

That single degree of separation (albeit tenuous) between myself and Walhachin, the Boer War even, is something to ponder. Growing up, my brothers had received the *Boys Own Annual* from their former Cub Master at Christmas and these were full of stories of the struggles between the British and the Afrikaners as the nineteenth century turned over to the twentieth. The names — Transvaal, Mafeking, Natal — entranced me. Robert Baden-Powell, the founder of the Scouting movement, had distinguished himself during the Boer War, which was probably the reason that my brothers received the books as gifts. But still that history had its resonance for a girl growing up in Victoria with its own traces of colonialism, its Majors at the Bengal Room gazing mournfully into their gin.

My mother scoured the Goodwill on lower Yates for riding breeches and boots for me, a worn trace of those soldiers. Tags on the jodhpurs might hark back to India or Jermyn Street and sometimes the aging leather boots had old-fashioned trees within them to keep them shapely. I imagined someone — maybe a widow or a landlady — collecting up all the old garments and putting them in a carton to be taken to the Goodwill, for who else but a girl whose allowance didn't stretch to proper riding clothes would want such things? They were impossibly cheap. $1.29, or $0.75. The detritus of lives passed, and now forgotten.

Of course it makes no difference that the house my Brownie pack entered to gaze upon the spoils of colonial hyper-confidence and activity was not the Newcombe house. I knew nothing of this then. I knew no First Nations people then. When we took a Sunday drive out past Brentwood Bay, we'd pass the tidy old farms on West Saanich Road, the fancier houses near the water, and then we'd come to the Reserve. Small noises would come from my parents' throats. They disapproved of the unpainted houses, the untidy yards, the dogs everywhere. It wasn't until much later that I learned anything of the history that allowed for such discrepancy between the communities. Paved sidewalks and prosperity on one side of the line, poverty on the other. Yet no one pointed out that each of those homes on the Reserve had a smokehouse behind it, that in spite of education policies that almost exterminated the cultures and languages of the original inhabitants of the coast, there was evidence of pride and dignity. Nobody mentioned or noticed that there was no need to clear out the wild plants in order to have gardens.

Gardens are an attempt to mirror Eden. But what if you already lived there? What if you could step out your door and pick huckleberries, salal, the new tips of thimbleberry to steam like celery? What if you could dig the roots of the blue camas to dry, springbank clover tasting like young peas, wild onions to flavour your stew? Or climb down to the beach to the clam beds, carefully terraced over the centuries. What if walking in the woods was like wandering through a vast and beloved place of abundance? Why clear the earth of all these life-giving plants in order to have…grass?

Once, riding my bike from a temporary residence out on Ardmore Drive to my summer job near Brentwood Bay, passing through the Reserve, a very old man came out to call off a dog that was lunging at me. "Here," he said, "try this," passing me a knife with something speared on its tip. And I ate a slice of warm salmon right out of the smokehouse. It tasted of the sea, and campfires on cool evenings, buttery and smoky. It was utterly of the place and time. And no one I knew was eating it. Except there.

In autumn of 2009, I was reading *Trees of Greater Victoria: A Heritage* and was completely surprised to come across this information:

> A rare heritage evergreen species, known as live oak, *Quercus virginiana*, at 144 Dallas Road, 36 inches (91 cm) in diameter, 22 feet (6.7 m) tall, has an amazing spread of over 50 feet (15.2 m). It is on the old homesite of C.F. Newcombe, outstanding Haida Indian authority for whom the Newcombe Auditorium at the Provincial Museum is named.[7]

So not a eucalyptus at all! Instead, it was a tree I'd read about in southern American literature, a tree associated with William Faulkner and Walt Whitman, a tree ancient and gnarled, draped with Spanish moss. In fact, the tree became a code for Whitman's robust homosexuality in the much-discussed "Live Oak, with Moss":

> I saw in Louisiana a live-oak growing,
> All alone stood it, and the moss hung down
> > from the branches,
> Without any companion it grew there,
> > glistening out with joyous leaves of
> > dark green.

But why did Newcombe choose a live oak, I wondered? Living in a city surrounded by wonderful native species — and himself a man who knew the value of plants and how a culture utilized them for medicines, commerce, and the practical business of everyday life — why choose a tree from the southeastern United States? Although common in southern cities, it is a tree happiest in parks, large estates, and on riverbanks where it can access damp sandy soil and where it can spread. The widest crown of any live oak is more than forty-five metres, belonging to a tree in Florida. I tried to figure out why that tree, in that place.

Next time I'm in Victoria, I drive over to Dallas Road (144 is next to 138; the lot was no doubt larger in the early twentieth century when Newcombe built here) and park across from Ogden Point breakwater. All those huge granite blocks were brought from Hardy Island, near where I live on the Sechelt Peninsula. I want to walk out on it as I did as a young girl with boyfriends on dark Friday nights. We'd pause to kiss as waves crashed against the exposed side. I always felt like I might fall — into the deep cold water of Juan de Fuca Strait or the more mysterious waters of human affection. Perhaps it was that fear of the deep that kept me from loving any of those young men, or even having them matter to me much, for I have a hard time remembering a single one of their names.

Instead, I pinch off another small branch of live oak with its deep green leaves, their undersides downy as a boy's face, and tuck it into my notebook, a small accordion pocket at the back provided for mementos. I don't see any of the acorns nestled in their deep cups, but this is a neighbourhood of squirrels, plentiful and industrious. The nameless branch at home is still pressed in a plant book, my efforts to identify it unsuccessful. Though now I can put a tiny bit of tape around its stem and write on it, *Quercus virginiana*.

No, this was no eucalyptus with its pinch of menthol, but a tree that grows quickly to a size large enough for shade, and is tolerant of salt-winds. One hundred thirty-eight Dallas Road is very exposed. There were fortified village sites along this part of the waterfront, beginning eight hundred to nine hundred years before contact, with moats and stockades; and it was kept shrub-free in order to encourage camas, Hooker's onion, and other food plants of the Lekwungen people. Later on the whole area was known as Beckley Farm, producing meat and vegetables for the Fort. Cattle and pigs ate the camas flowers before they could seed and the pigs even dug up the bulbs, eager for their sweetness. Little by little, exotic and introduced species took over from the delicate wild grasses and herbs; a live oak in soil once nurturing Garry oaks or arbutus. Yet Victoria's mild climate has encouraged gardeners to plant trees as exotic as bananas and palms so perhaps I shouldn't have been surprised to learn the tree's true name.

As a child I was impressed that Beacon Hill Park featured the tallest totem pole in the world, carved by Mungo Martin. It was erected in 1958 and is 127 feet tall. Sometimes I'd ride my bike over to the park and sit at the base of the pole. I could see the Juan de Fuca Strait and often there was a cold wind coming off it. I remember leaning my head back and trying to read the story in the totem's elements. No one talked about its imagery. To refresh my memory, I've checked the newspaper reporting of the time and find no mention of the symbolic elements of the work. Its size was emphasized, the method of placing it in a skirt like a huge candleholder so that guy wires weren't required.

The cedar tree that became the tallest pole grew at Muir Creek off the West Coast Road near Sooke. Brought to Thunderbird Park, it became graphic with Mungo Martin's family stories. Beginning with Geeksan wrapped in a blanket at the bottom, followed by the cannibal bird Huxwhukw, the crest animals rise one by one — killer whale with its formidable teeth, sea lion, eagle, sea otter clutching a fish, another whale, beaver, a man, seal, wolf, crowned by three men, two of them wearing blankets against the chill winds off Juan de Fuca Strait. And perhaps against loneliness. Beacon Hill Park is far from Fort Rupert on the northern edge of Vancouver Island where the stories had their origins.

I saw the world as an animated place. Walking the beach in search of bark for our stove on Eberts Street, I'd find long lizards of root wood tangled in amongst the logs with the faces of ravens peering out of the grain, ovoid knots forming the eyes. The monkey-puzzle trees with their serpent-like branches dropped cones scaled as the garter snakes my brothers would drop down my T-shirt, leaving me frozen in horror as the dry terrified animals slithered out above the waistband of my shorts. It wasn't that I was frightened of the snakes themselves. I could spend ages looking at one sleeping on Moss Rocks, even touching the pattern on the scales with a tentative finger. But feeling them against my bare skin was enough to make me pee my pants. Which I suppose was the idea. I learned to pretend indifference as I got older, which meant that I lost the knowledge of reptile skin against my own.

I saw the tired heads of elk in the bare branches of Garry oak and black bears nosed their way from the burned wood of old bonfires. So looking up into the faces of Mungo Martin's crest animals staring steadfastly out to sea from their perch on the tallest totem was akin to reading a story from a culture adjacent to my own but which shared elements, a sense of the numinous, and to recognize the familiar amidst the strange. These were not my crest animals, but they were part of the landscape I loved, and some of them had formed me as surely as they had formed any child born into that locus.

The squeak of the wooden wheels of change, carts dragged by sleepy oxen to and from the Fort, bringing turnips, potatoes, slabs of bloody meat.

Squeaking as they brought food to the pesthouse where a five-year-old child was taken from the Prince Alfred in 1872, "the house formerly occupied by Mrs. Nias on Beckley Farm, which is now Government property and which was fitted up yesterday for the reception of the little sufferer. The yellow flag waves over the house."[8]

Ploughs turning and opening, carts waiting to take away the stones, gulls wheeling in their wake in the clear blue skies over the fields of Beckley Farm. Imagine Lekwungen families watching as their own cultivated patches of common camas and great camas disappeared without consultation. A child in the shadow of a Garry oak, digging stick in hand, puzzled by the sad turning away of a parent while the pigs rooted and feasted on broken bulbs turned up by the plough.

The smallpox patient taken to the pesthouse died and was buried somewhere on the waterfront, marked now with a small plaque, though the exact location of her grave has been long forgotten. That little sufferer, bones under the foreign grasses and Scotch broom of Holland Point.

Charles Newcombe was an intrepid collector, not just for the Royal Botanic Gardens, Kew, in Richmond, near London, but also for the Field Museum in Chicago and other institutions all over North America and abroad. He wanted Canadian institutions to recognize the value of the materials but had only intermittent luck for years, though he received encouragement from George Mercer Dawson of the Canadian Geological Survey. It became a deplorable competition, boats going up and down the coast carrying fervent hunters for the authentic object. Poles, bentwood boxes, rattles, masks, even something called "osteological collections"—in other words, skeletons. Some museums wanted characteristic materials from the discrete cultural areas. Some went for quantity over quality. Some were generous with funding and urged their agents to go to extreme measures to procure trophies of special interest.

The missionaries were involved. At first they encouraged their converts in the Native villages to burn totem poles and other aspects of traditional life. But then they realized how valuable these items were. In Jan Hare and Jean Barman's *Good Intentions Gone Awry: Emma Crosby and the Methodist Mission on the Northwest Coast*, there is a telling passage in a letter home from Emma Crosby (wife of Methodist missionary Thomas Crosby) to her mother. It is 1875, and the Crosbies have been visited by James Swan, erstwhile Indian agent from Washington and agent for the Smithsonian Institute who is also collecting for the upcoming Centennial Exhibition:

> Another stray str. visited us a few weeks ago & a U.S. revenue cutter having on board a U.S. Indian agent looking up Indian curiosities for the approaching Centennial Exhibition. He took dinner with us & I should say was a fair specimen of the U.S. Indian agent as reputed to be — not inconveniently high-minded.[9]

This visit was further elaborated upon in Douglas Cole's *Captured Heritage*:

> Inducing the Indians to give up their heathen ways, Methodist missionary Thomas Crosby had persuaded many to remove poles from outside their houses, and, though some of these had been burned, others were collected "in a sort of museum." Swan bought one, a finely-carved forty-foot specimen, and he hoped for more. Both Crosby and C. E. Morrison, the HBC trader, agreed to gather a collection for Swan which they would send to Victoria.[10]

Realizing how potentially lucrative the objects were, Crosby began to collect for himself as well and was ideally placed to persuade those under his pastorate to give up their cultural wealth.

There were rivalries to see which collectors could acquire the most poles, whole houses, coppers, and feast dishes. On one collecting trip for Franz Boas, Newcombe sailed to Ninstints, a Haida village off the tip of Moresby Island in the Queen Charlotte Islands. Chief Ninstints himself accompanied Newcombe as pilot. Newcombe reported to Boas that the Chief "is half-blind and I hope to acquire many interesting things in that almost wholly deserted village for you." (The population of Ninstints had been decimated by smallpox earlier in the century.) One of the crew diverted the Chief's attentions while a few "osteological" items were located, though not what had been hoped for. However, other village sites were visited and mortuary areas raided while the "half-blind" Chief Ninstints was kept busy in other ways.[11]

Yet the collection at Kew takes us from grave-robber and villain (of a sort:

these things are never black and white and at the time Newcombe was in hot pursuit of artefacts, it was an acceptable practice) to a passionate advocate for the importance of ethnobotany. A beautiful set of gambling sticks carved of Pacific crabapple is as lovely as anything sold in a contemporary craft shop. And a halibut hook, formed of western yew and strong enough to hoist the weight of one of those bottom-dwelling sea denizens weighing in excess of 180 kilograms, is graceful and practical. There is some suggestion that collections such as these work to educate those whose cultures have lost their traditional practices. The carefully wrought baskets and implements hold the lessons of their makers — the way their hands worked the fibres, the marks of the adze on the wood — and do much to help us recover the past.

The Royal British Columbia Museum holds hundreds of photographs taken by Newcombe showing village sites in all their intact dignity, houses gazing solemnly to sea with canoes pulled up in front and a few people purposefully digging clams or sitting on logs in sunlight. For all that he removed from those places, Newcombe preserved a complex codex of place and culture in that photographic record.

A natural historian like Charles Newcombe would surely have known about the useful qualities of *Quercus virginiana*. Is this why he planted one? He would have known, for instance, that it was an important tree for shipbuilding (the US Navy had large tracts of live oak for this very purpose), the massive arching limbs finding their way to the ribs and knees of ships, the wood itself heavy, strong, and hard.

The Houma people who lived in what is now Louisiana cherished the live oak. After their own language was absorbed into French, they called it *chêne vert* or the green oak. Its bark provided red paint for the post on the Mississippi River that marked their hunting territory in what is now Baton Rouge (or red post). The acorns were an integral food supply; live oak acorns being high on the list of palatability, they were pressed for their oil and were soaked (to leach out the tannins), dried, and ground to make meal.

The anthropologist John Swanton wrote extensively on the Aboriginal peoples of two geographical areas in his long career — the Pacific Northwest and the Southeastern United States. Charles Newcombe met Swanton at Skidegate in 1900, when Newcombe was on a collecting trip for Stewart Culin and the University of Pennsylvania. The two men got along quite well.[12] After his work with the Haida and Tlingit, Swanton went on to study the Muskogean peoples of the Southeastern United States (the Houma are included in this linguistic group).

It's a fanciful stretch, but I like to think that maybe Swanton inspired Newcombe to plant his live oak when the latter built his house on Dallas Road. Maybe he even sent a root, an acorn, a small sapling, knowing that the species could withstand the salt-laden wind off the Juan de Fuca Strait. Potatoes, turnips, and the broom which so drastically changed the face of Victoria, graveposts and houseposts and the entire regalia for winter ceremonials travelling by train across America to the Chicago World's Fair in 1893, gambling sticks of crabapple wood destined for the Royal Botanical Gardens, Kew... In such ways cultural and botanical knowledge travel the well-worn roads and rivers of human experience, participating in exchange, small acts of mercy and theft, and larger ones of kindness and exploitation.

The Salvage Paradigm
I read whatever I can about Wawadit'la. The late Wilson Duff, the anthropology curator at the Provincial Museum who encouraged the construction of Wawadit'la, said:

> This is an authentic replica of a Kwakiutl house of the nineteenth century. More exactly, it is Mungo Martin's house, bearing on its

houseposts some of the hereditary crests of his family. This is a copy of a house built at Fort Rupert about a century ago by a chief whose position and name Mungo Martin had inherited and assumed — Nakap'ankam.[13]

According to Wilson Duff, Wawadit'la was one of two names that Martin had the right to choose from for his creation; it means "he orders them to come inside." Yet it seems there was no single house on which Mungo Martin modelled his Thunderbird Park version. Rather, there are several.

The house where Mungo Martin was born was the home of his mother's uncle, the old chief Nakap'ankam.[14] It appears this house was never finished — it had no frontal painting. In what seems to be an homage to his father, who had come from Gwayasdums on Gilford Island, Martin uses an image from a house there for the frontal painting of the Thunderbird Park Wawadit'la, though with some stylistic changes. Martin reverses figures on the houseposts from those on Nakap'ankam's house, and synthesizes other design elements. In a paper given at the 1987 annual general meeting of the American Anthropological Association in Chicago, Ira Jacknis considers the various sources informing Mungo Martin's concept for Wawadit'la and the way the house both reflects the ideal and the pragmatic in terms of its influences: "the non-native context of Martin's house may have encouraged the artist to synthesize diverse Kwakiutl forms into a kind of 'super-artifact.'"[15]

And what is a copy, what is authentic? Can the house of a Kwakwaka'wakw chief, built in Lekwungen territory on southern Vancouver Island, far from the village of its origins, built with the sense that everything it represented must be commemorated because so much of that culture was disappearing, can this house truly be called a copy? Houses, like songs and stories and other aspects of culture, exist across time and place, in a moment that is ever-present. When I saw that old man, who might have been Mungo Martin, working in Thunderbird Park, it did not occur to me that he had not always worked there, had not always lived in the house with some sort of sea creature painted on its facade. A child leaning her blue bike against a rock and watching was taken into that moment. The ravens in the trees knew this, their vocabulary unchanged over the centuries, in a city where the bodies of Native children and the young passenger from the *Prince Alfred* who died in the pesthouse at Holland Point lie under the ground in proximity. Charred stones from the pits where springbank clover was steamed can be found in the sand. The buried streams remember their routes to the Inner Harbour, under what's now the Empress Hotel, and where the old lodges at the original Songhees village site, though vanished, still give off the faint scent of cedar if the wind is right.

Anthropologists might disagree about what is authentic and what isn't, made anxious by preoccupations of contact and culture and the "salvage paradigm."[16] This isn't surprising: careers are built on fine distinctions. There is evidence that Mungo Martin felt that he was the last of his kind in some respects, working against time and oblivion. Wilson Duff wrote to a friend, "Mungo is convinced: (a) that this will be the last 'house-warming' potlatch and (b) that nobody else but him knows exactly how to do the whole thing properly."[17] This house recreated and adapted and given new meaning, then: a repository of both memory and history, those duelling conspirators.

What is particularly interesting is that anthropologists can engage in considerations of authenticity so many years after the plundering of the Kwakwaka'wakw villages and the loss of so much of their material culture, in part because Charles Newcombe took hundreds of precise photographs documenting the villages. One can look at them, a single degree of separation, and approach something of the experience of gliding onto the beach at Kalokwis on Turnour Island among the canoes where the housefronts stare out to sea, their imagery and context intact. Or walk up to the group of people standing in front of Kwaksistala, a house on Harbledown Island, in 1900, children and adults wrapped in blankets, a few of them in headdresses. That house's sculpin front informed, in memory, some of Mungo Martin's work in Thunderbird Park, as of course did Gwayasdums. We can almost remember, looking at these photographs; almost trace the trajectory of the artist's work back to his original home at Fort Rupert on the northeast coast of Vancouver Island, where clams were dried by the fire and elegant hooks of western yew might bring up a halibut. We can almost stand there in our otherness, our clothing slowly absorbing the smell of cedar smoke and salt.

In 1906, Charles Newcombe sold a portion of his private collection of artefacts to the Canadian Geological Survey for $6,500; this paid for the house he built the next year at Ogden Point, the house I'd hoped was the one I'd visited as a Brownie, the same house now sheltering paroled inmates from federal prisons. It no longer even carries the Newcombe name but rather that of a deceased chaplain of one of the local penitentiaries, its original owner forgotten by the neighbours — who maybe even wonder why the large tree on the corner of their property is protected by heritage status and can't be cut down.

Newcombe increasingly devoted his attentions to the British Columbia Museum of Natural History and Anthropology, which had opened in 1887 and had concentrated, until then, on stuffed animal specimens under the direction of its curator John Fannin rather than the cultural riches of the province's Indigenous peoples. If this omission had resulted in the materials being left in villages of their origins, then one could commend Mr. Fannin from this century to his own. But alas, as a visitor to New York observed in 1900, there is "a veritable forest of totem poles" at the American Museum of Natural History in Central Park.[18] And the Field Museum in Chicago, through the collecting tenacity of George Dorsey and others, had the contents of many Haida burial caves and gravehouses.

Rising to the heights of umbrage, *The British Colonist* featured this headline in November 1903: "LOSS TO BRITISH COLUMBIA Through The Depredations Of United States And Other Foreign Collectors Of The Province's Most Valuable Indian Relics." The article goes on to say:

> The government might take a leaf out of the book of the universities of the United States. Several of these have collectors constantly in the field collecting and cataloguing relics and collating their histories. The province could do likewise and appoint some one [sic] to conserve what rightfully belongs to us.[19]

Of course, the matter of ownership goes uncommented upon. Newcombe continued to do some collecting as well as writing and publishing on historical and zoological topics.

Memory provides a curious and unreliable template. In an attempt to fit a house into a specific memory, I found another house, its gracious rooms and verandah given over to the healing of men released from prison and trying to find their way back into the world. I discovered that another house I had known in my childhood and had believed then to have stood in Thunderbird Park since the beginnings of time (whatever that meant to me then) contains a series of paradoxes, both territorial and cultural. Nothing is as permanent as change and the shifting boundaries of how we remember the past. A house is more than those who live in it, its secrets encoded in its architecture and domestic history long after its residents depart this earth.

That child on her small blue bicycle explored the fringes of a time and place on the cusp of change, though she didn't know it then. Stepping into the home of a missionary, it didn't occur to her that the masks on the wall and the rattles used by a shaman somewhere on the coast of the province had been gathered improperly. She had no idea the place where she lived had been colonized so thoroughly that even the namesake plant of her own playground at Clover Point had been supplanted by invader species. Yet she grew up with such clear maps in her mind of that beloved terrain—the snowberry bushes of Lover's Lane hung with the treacherous nests of wasps, the location of the beautiful fawn lilies on Moss Rocks that turned their faces to the world after pollination, a small park where curls of cedar drifted to the ground to be collected by children — that as an adult, sleeping far from those familiar streets, her dreams were often filled with their houses and their trees, the waves washing onto the shores of Ross Bay a distant sacred music.

Olea europaea
Young Woman with Eros on her Shoulder

> Full noon, July...
> If the olive groves didn't exist
> I would have invented them...
> — Odysseas Elytis, *Eros, Eros, Eros: Selected and Last Poems*

I'd treated myself to a cabin on the ferry from Brindisi to Piraeus. My journal tells me it cost about thirty dollars, which was expensive in 1976; but I'd come from Madrid by train across southern France and hadn't slept for three days or nights. (One day I will tell the story of being alone in Madrid without luggage, which had been lost en route from Canada. It was my first trip, I was twenty-one years old, and I still don't understand why I didn't turn around and go home.) The compartments were full and the seats cramped. I'd doze off for a few minutes and then someone would inadvertently elbow me as he or she reached up for baggage or else rearranged clothing or rummaged in a basket for a leg of chicken or bottle of wine. I kept pinching myself to remind myself I was in Europe — that, and everything else kept me awake.

The cabin was plain but comfortable and after leaving the party of Greeks I'd met on the train from Rome to Brindisi—a group of guys heading home from jobs in England to a family wedding in Athens—I crawled between clean sheets and slept. Waking, I stepped outside my cabin just as the ferry was passing through the canal cutting across the Isthmus of Corinth.

I'd never seen a blue like that of the water and sky. Every adjective was called into being; and every one found wanting. Above the city of Corinth, on the Peloponnese, was the Acrocorinth with its remains of temples and fortifications. There was an important sanctuary to Demeter and Kore there; I'd studied their cult and it was glorious to be so near a site associated with two faces of womanhood—the mother and the maiden. The sun, even early, was warm.

My Greek friends came by to offer me breakfast on the deck of the ferry—they pulled cheese and apples from their rucksacks (they'd slept on the deck, under the stars), provided small slugs from a flask of Metaxa, and little cups of coffee they brought from the ship's cafeteria. The previous night they'd surprised me by quoting poetry—Yannis Ritsos—and I asked what poems they had for a morning sail through the Isthmus of Corinth:

> The harbor is old, I can't wait any longer
> for the friend who left for the island of pine trees
> for the friend who left for the island of plane trees...[1]

I recognized Seferis. "Now say it in Greek," I asked. One of them continued the poem:

> Τ' άστρα της νύχτας με γυρίζουν στην προσδοκία
> του Οδυσσέα για τους νεκρούς μες στ' ασφοδίλια

("The night's stars take me back to the anticipation / of Odysseus waiting for the dead among the asphodels...").[2] I wanted such ease of transport, between

London and Athens, Greek and English, those days of Odysseus among the asphodels and myself on the deck of a ferry sailing for Piraeus.

Travelling through Spain and France, I'd grown to love the sight of olive trees. How beautiful they were, the gnarled trunks of the old ones, and the grey-green leaves trembling as the train passed. I knew from my Classical Studies courses that olive trees were sacred to Athena, that the first olive tree was a result of her striking her spear into the soil of the area which became her city, Athens, and that the original tree still grew there.

I'd also grown to love olive oil in my very early twenties, which is the time I am writing of — though it was something I'd never had at home. Reading Elizabeth David and Jane Grigson had made me realize that a world of food waited, just as a world of other wonders did, and it made me impatient for my life to begin. That was one reason I was heading to Crete. I was hopeful that all the things I'd wanted and wished for in Victoria and which hadn't materialized — love (my friends were all pairing up and the men I wanted never noticed me), a table laden with figs, wine and fine cheeses, herbs picked from the land itself, a sense of myself as a writer in my own right, with passion and purpose — were possible here.

Who knows why the young attach such yearning to places other than home, and take to the skies, the seas, the mountains of the world? In those years (and I dare say in these years as well), they were abroad with their rucksacks, their passports, clutching well-worn copies of *Europe on Ten Dollars a Day*, watched by those who'd done the whole thing a decade earlier on five dollars a day. I had none of those books — only Henry Miller's *The Colossus of Maroussi*, Lawrence Durrell's *Reflections on a Marine Venus*, and my beloved copy of *The Odyssey*, in Robert Fitzgerald's wonderful translation. In these books, there were olive trees, and lamb bathed with oil grilling on open fires. I thought I could travel the way my heroes did, with luck and the blessing of the gods.

Two Americans began a conversation with me; they wanted to know where I was going. After a private consultation with each other and their guidebook, they wondered if they could tag along to Crete. My Greek friends helped us to figure out the boat connection to Herakleion. This gave us a day in Piraeus. The harbour was filled with boats and sailors. Catcalls followed us everywhere, and hisses—even into the *taverna* where we went for a meal. A carafe of retsina materialized on our table with a table of men on the other side of the *taverna* bowing in our direction.

It was a night ferry, and most of its passengers were Greek, some carrying cages of chickens; one man had a goat with rope around its neck. There were bunks with mattresses covered with cracked vinyl, wide enough for two. The three of us shared one bunk after a meal of macaroni with nutmeg-flavoured sauce, washed down with a glass of rough red wine. I can't recall the names of the young Americans, but I remember the woman and I settled lengthwise on the bunk and her partner, a man, stretched out across our feet and wrote in a journal. All around us people talked, drank, and smoked cigarettes that filled the compartment with grey smoke. Children cried; the goat bleated and was soothed by its owner. The smell was powerful—of cheese, and sweat, and wool. Someone played a bouzouki and I fell asleep to its strange familiar music.

I don't know what I expected of Crete but it surpassed anything I had imagined. To approach by sea, at dawn, the water dark and Herakleion glowing in the first light. To stumble off the ferry and into the streets where people offered rooms, transportation in dusty taxis, guide services to Knossos or the archaeological museum.

We found the American Express office and cashed traveller's cheques before finding out when the bus left for Agia Galini. Tickets purchased and packs stowed by the bus while the driver went out for a meal, we went in search of food for the bus trip. A bakery offered loaves of chewy bread; a small store had bins of tangy cheese, chunks of which were weighed and placed in plastic bags; and I chose some olives—bright green, darker green, and cracked, purple. Tomatoes. I also bought a tub of thick yogurt. The woman in the store asked, "*Meli? Meli?*" She took a spoon and dipped it into a bucket, holding it up to my mouth. "*Meli*," she smiled. I tasted. Ah, honey! And it became one of my first Greek words — μέλι. And yogurt with this honey — γιαούρτι με μέλι — became my preferred breakfast, the sheep's yogurt thick and creamy, honey ribboned through it like gold.

A revelation: the underground public toilet near the bus station. One descended stairs to a cave, where a very old woman dressed in black guarded the entrance. Her anteroom held a bed, a table, and a chair. An ikon hung above the bed and a spidery piece of knitting rested on the table. The woman held out her hand, preventing me from going through to the poorly lit room beyond, and I finally realized that the basket beside her held sections of toilet paper. One had to pay for a few squares before entering. The odour was appalling. There was no door to close. And the toilet itself…wasn't. It was a hole in the floor, with two slightly raised metal rests for the feet beside it. There was nothing to hold on to. I squatted carefully and peed into the hole. The paper was slippery, waxy. I don't know if I'm correct in remembering that there was nowhere to wash my hands afterwards. The experience felt strangely mythic, and yet I wasn't sure what it portended.

When we finally boarded the bus, it was as though we were stepping into a small shrine. The dashboard held many objects, ikons and votives, with still more hung above the window. Some were fringed with small bells, and when the driver turned a corner, they rang and chimed. The driver mostly drove with one hand while the other worried at a string of beads on his wrist. Music filled the bus, mournful at first and then beautiful as it entered my blood, finding my pulse.

As we left Herakleion, the road became rougher and climbed higher. When I looked around me, I realized that all the men, young and old, were wearing beads on their wrists and the clicking I heard, along with the bells, was the sound of them as we rounded corners, passing sheer drop-offs without any abutments to prevent the bus going over the cliffs. Tiny churches appeared here and there on the high slopes, and every time we passed one, the beads clicked as people crossed themselves in the Orthodox fashion. A man riding sideways on a donkey stopped as the bus hurtled by and that was my first sight of the old Crete—baggy black trousers tucked into high boots, an elaborate head wrapping, long moustaches drooping from the mouth. And there were mountains everywhere, rocky and pierced with caves. In one of them, I knew, Zeus had been born—which didn't surprise me. It was an island of beginnings, somehow. I hoped for one myself—a new beginning.

Tell what you saw as the bus raced towards Agia Galini. Mountains. Tiny villages, white with churches, perched so high that you couldn't imagine a reason for a village to be there until you remembered the history of Turks, Venetians, Germans; and yet a road lurched up the mountainside, sheep and goats grazed on the stony slope, a few donkeys carried their riders uphill, laden with sticks and sacks.

Steep rises covered in dittany, juniper, plane trees in the squares of the towns we passed through, shady and green. Holm oak and kermes oak. Rocks covered in low-growing thyme—by now you'd eaten your γιαούρτι με μέλι and knew that the bees had worked the thyme flowers to create this ambrosia, the first honey for which you'd been able to detect its origins.

Groves of olives, and long rows of grapes on the fertile plains. Through the open windows you smelled dust and unfamiliar wind.

Signs in Greek, which you yearned to be able to read. (You had your grammar and were trying to master the alphabet.)

You passed houses almost smothered in vines, the window frames and doors painted blue. Rusty oil cans held geraniums and lush basil tomato plants climbed the whitewashed walls. Old women shrouded in black sat on chairs and held up a hand to the driver of the bus.

Outside a church in a small town, you saw an Orthodox priest eating an apple.

Thirty-three years later, I am thinking about that girl on the bus, heading to the unknown. I am thinking about her innocence and her hope. I am in the Montreal Museum of Fine Arts when I find myself stopped in my tracks by a small terracotta sculpture.

It's not much more than thirty centimetres tall. The label tells me it's from the first half of the second century BC, possibly from Taranta, Italy. The label says it's a young woman with Eros on her shoulder from an *ephedrismos* group. I have no idea what an *ephedrismos* group is, but I note down the information into my notebook and move along to the rest of the Mediterranean archaeology exhibit.

And yet she stays with me, the young woman. Eros holds out both hands, as though in benediction and she looks up at him over her right shoulder. Her right breast is exposed, the draperies of her chiton show that she is using the muscles of her thighs to support the weight of the god.

I try to find out something about *ephedrismos*. It turns out it was a game played by children in ancient Greece. Julius Pollux tells us: "They put down a stone and throw at it from a distance with balls or pebbles. The one who fails to overturn the stone carries the other, having his eyes blindfolded by the rider, until, if he does not go astray, he reaches the stone, which is called a dioros..."[3]

There are other examples of *ephedrismos* figures as well as depictions of players on vessels, even the special jars used to hold oil for funerary purposes. Several are of girls partnering each other, and one fascinating rendering, from a fifth-century BC funerary flask, shows a satyr with a *maenad* on his back. These are both followers of the god Dionysus, associated with wine and sexual frenzy, and there is something deliciously suggestive about the notion of them playing a child's game in a context that is decidedly adult.

Back to the little terracotta figure from the Montreal museum: I wonder why Eros is this young woman's partner, and why his hands are raised, not covering the young woman's eyes as the game's rules require? I know that Eros is sometimes represented as a child god, playful and mischievous. In his ikonography, he is most often shown with a bow and arrow, ready to shoot desire into the hearts of his victims. Yet here he is a participant in a children's game, riding the shoulders of a young woman whose breast is exposed, her cheek resting against his abdomen, close to his groin.

The driver drove very fast down the hill and stopped in a square with a planter of flowering hibiscus in its centre. This was Agia Galini, Saint Serenity, a village right on the edge of the Libyan Sea. There were some restaurants around the square and boats at the quay, with more of them pulled up to shore.

I walked up through the village with the Americans. They paused by a house with a Rooms sign and knocked on the door. I kept walking, almost to the outskirts of the village, where I'd noticed a house as we drove in. It was across from the church and its separate bell tower and had flowers tumbling from cans and pots all around the front door. I don't know why it caught my eye, but a room was waiting for me there in Angela's house, painted blue with green trim; impossibly cheap, with a bed, a table, and a few hooks on the wall for my clothes. "Come in, come in," Angela said, and brought a tray with a glass of water, a spoon, and a jar of cherry preserves. She took my pack from my shoulders. I could have been a goddess in disguise, not a potential tenant at all. The sweet taste of cherries on the tongue, and then the long fall of cool water down my throat.

Most of this from memory, by heart. I have a journal with the kind of entries a young woman of twenty-one would make, eyes open, her finger on her own pulse — the colours of the earth on a hike up above the village; a funny anecdote about a German woman marching up to a *taverna* owner to ask that he turn down the music (that heady delightful stuff that poured out of every café or tavern) and leaving in a huff when he refused to; loneliness; the sight of a loom on a rock floor, strung for weaving; a donkey saddle leaning against a wall, broken. I remember the bakery, also across the road, with its fierce wood-burning ovens, and how most women in the village took their casseroles there to bake — most didn't have ovens in their homes — and how on a bread day, I'd buy a warm brown loaf, pulling chunks off to eat with cheese on my walks up to the olive grove above my house.

Angela and her husband grew olives, oranges, lemons, and grapes, as well as vegetables and herbs. Some days when I went into my room, there would be a bowl of sliced tomatoes and cucumbers, drizzled with oil from the village press, where each family took in their fruit and participated in the pressing. On the days when my sheets were changed, I'd find my nightdress folded on the pillow, smelling of the myrtle bushes where Angela dried her linens, and I'd realize she'd taken it out with her own laundry.

From memory, by heart: the clicking of beads as men watched soccer in the bar with a television. The mournful bray of a donkey at dawn. A woman delivering fresh cheeses to the store where I bought my yogurt and honey, and the taste of that cheese — goaty and faintly pine-flavoured, as though the animal had been feeding on rosemary, which was entirely possible, as it grew everywhere, loud with bees. Dancing at the *tavernas* in the evenings after golden retsina while travelling musicians played their eerie, repetitive songs into the small hours.

There were two villages. There was the one with tourists (and I count myself among them), sitting at the sidewalk tables and talking, laughing at all hours of the day, or else heading over to the bay just beyond the last house, towels over their shoulders. We loved Joni Mitchell, and would listen to *Blue* at night while the lightning flashed from cloud to cloud and the donkeys brayed. And there was the village that went on as though none of us were there, or mattered. The narrow streets with closed doors and shutters. Storekeepers completely indifferent to the impatience of a young person wanting to pay for a slice of baklava to take to the beach and going on with the job of stock-taking or arranging loaves of bread in a basket by the door. The men who came down to a bar to drink raki and play backgammon, their worn trousers stuffed into boots. The women, secretive, dressed entirely in black, sweeping a few chickens into a doorway with a handful of straw.

A very old man, a fisherman with a bright blue boat, used to bring me slices of melon when I sat at the dock and read my book. One day he brought his son, whom I will call Agamemnon. He was older, had served in the army, and spoke English only marginally better than my Greek. He owned a *taverna* where I'd eaten a couple of times — stuffed tomatoes made by his mother, Maria, salads piled with salty cheese and thick onions, bottles of cold beer. We walked out a few times in daylight, walked to the end of the long pier and back; he pointed to birds, the distant horizon, a boat rigged with sails. His hand, when he reached to hold mine, was calloused, from helping his father. Mornings, he met his father returning from a night of fishing, lifting off boxes of strangely coloured and whiskered fish, untangling nets, removing shells and bits of kelp. His father always pulled a flask from his pocket and offered it around. No one wiped the rim first; drinking from it was intimate as kissing. The burning in my throat told me it was raki. I remember the feeling of the salt-stiffened ropes as we hauled in the boat to the shore, my own hands rough and bleeding afterwards.

"If you deconstruct Greece," wrote Odysseas Elytis, "you will in the end see an olive tree, a grapevine, and a boat remain. That is: with as much, you reconstruct her."[4]

When Angela brought me a dish of cucumbers drizzled with oil, she asked, "Do you like?" And yes, I said I did, very much. She woke me one morning to go with the family to harvest their olives. They did this over a period, taking the fruit that was about three-quarters ripe. This made the best oil. Let them ripen more, I eventually understood she was telling me, and the oil will not be good.

Angela's family grove was not large but very beautiful; the trees well tended and growing on a slope facing the sea, about a kilometre above the village. The donkey came with us, worn panniers on his back. The panniers carried lunch on the way up and olives at day's end. There was Angela, her husband Yianni, and their older daughter Eleni (oh lovely Helen, who was engaged to Demitreos, doing his military service; she looked like a Byzantine ikon); their younger children were in school.

They spread wide, fine-meshed netting under the trees, three or four at a time. There were long sticks with straw at the ends, like brooms. The idea was to agitate the branches so the riper olives fell. The straw was dragged along the branches like a fork, loosening olives that didn't respond to beating. Then the olives were gathered and sorted, leaves and twigs brushed away, overripe ones saved for the animals.

After a few hours of this, we ate hard-boiled eggs from the family's chickens, with ripe tomatoes and bread. They also ate onions, crunching into them like apples, but this didn't appeal to me. Rough red wine was shared from a single bottle.

When we'd finished for the day, Yianni took the donkey down to the building where the olive press was. It was important that the olives be pressed immediately. The family would receive a substantial portion of the oil, but some went to the co-op to pay for the cost of the press and its upkeep.

I loved being in their grove, the wind rustling through the grey leaves, the grass dry and fragrant. There was pungent sage and *rigani*, thyme and chamomile, and I could see the seedpods of poppies left from the spring. "Watch for snakes," they warned, and I nervously kept an eye out for movement.

Several million olive trees grow on Crete, some of them more than one thousand years old, with trunks measuring twenty metres in circumference. They have the capacity to sucker strongly from severed trunks so some trees could be far older than their present form might suggest—and Greeks will tell you seriously that all olives come from rooted cuttings taken from Athena's original tree. Fossil olives and olive wood found on Thera indicate production predating the great volcanic eruption of roughly 1600 BC, which destroyed the Minoan civilization flourishing on Crete and exported elsewhere by trade and perhaps colonization. Frescoes from the early Minoan period thirty-six hundred years ago at the palace of Knossos at Herakleion show olive trees, the long leaves as lovely as any growing now.

I took the bus to Herakleion several times to visit the Museum there. The first time I went, I walked out to Knossos. I wanted to see that ancient palace and have an idea of its size, its shape, and function before I looked at objects from the site. It was very imposing — its throne room, shrines, courts, and the vast system of storage rooms. I wasn't sure about the vivid colours used on some of the reconstructed areas, or how much of the wall painting imagery to believe was original.

I know that many find the whole Arthur Evans enterprise problematic — a wealthy Victorian amateur archaeologist, he bought the land on which the ruins of Knossos stood and began his excavations in 1900 — and modern archaeological principles are certainly much more sensitive to authenticity and accurate interpretation. Still, beyond the painted pillars and patched frescoes, there was the undeniable sense of deep history. Closing my eyes, I could imagine its busy life — the potters, the metalworkers, those preparing food. The Minoan deities were largely female; the priestesses and goddesses in graphic representation held snakes or animals; their breasts were proudly bared to emphasize fertility, and there were often poppies nearby.

The Museum was wonderful, chock full of glorious, animated Minoan ceramics — octopi wrapped around jugs, flowers (the lilies were particularly lovely), marine life given an airy and naturalistic place of honour. The Minoans were a trading culture that excelled at metalwork, importing copper from Cyprus to alloy with local tin to make bronze tools, implements, and statuary, as well as weapons. I loved the gold jewellery, resplendent with bulls and bees.

But it was the frescoes that impressed me most. Their composition was so harmonious, space organized the way a composer might notate music, main theme embellished beautifully by use of colour and motif. My favourite showed a man stretching out to gather the long stigmas of saffron from crocuses scattered over the surface of the fresco, with long, undulating horizonal bushes containing the activity. (Years later I was startled to read that this particular work had been erroneously restored by Arthur Evans because contemporary forensic methods show that the man was in fact a blue monkey gathering saffron!) *Crocus sativus* could be found all over Crete, and so it was a moment when the past transected perfectly with the present.

Giant amphorae held olive oil and wine, two constants of Mediterranean culture. The plants providing these important resources were evident in the art, on the seal stones so beautifully carved. I saw slender leaves incised into stone, little fruits dangling. Some of the small seals were so worn that all I could see was a cup of wine raised to a mouth.

Later on during my time on Crete, I visited Phaistos, and Hagia Triada. In many ways, I preferred the wilder Phaistos to Knossos. Beautifully situated on a low hill west of the Mesara Plain, it had none of Arthus Evans's fanciful reconstructions. "Phaestos [*sic*] contains all the elements of the heart," wrote Henry Miller after his experience there in 1939, describing so beautifully its mythic quality in his great book, *The Colossus of Maroussi*, first published in 1941. He explored the site in the company of its caretaker, Kyrios Alexandros, whose son held the same position as his father had when I came, years after Miller. The son never treated me to Mavrodaphne wine, however, and I remembered Henry Miller's description with envy: "He opened a bottle of black wine, a heady, molten wine that situated us immediately in the centre of the universe with a few olives, some ham and cheese."[5] As a result, he'd felt closer to the sky than ever before, a feeling I also had at Phaistos, where the blue dome met the earth in an expression of physical love.

The palace was itself compact, of a piece, somehow, devoid of the erratic sprawling organization of Knossos, which spread itself over a large chunk of land near Herakleion. It's thought that the villa at Hagia Triada, about three kilometres from Phaistos (two elegant wings flying out from a central courtyard and surrounded by a verdant valley), was a summer residence for the priest-king of Phaistos. And there's also another ancient site, Gortyn, on the plain of Mesara, which is where Zeus took his abducted paramour Europa (some say he took her from Lebanon) and made love to her there under a shady plane tree, impregnating her with triplets—Minos, Rhadamantys, and Sarpedon, all of whom became kings of Minoan palaces.

These stories hung in the air like golden dust. You could believe them or not, but you breathed them in regardless. The profiles of the priestesses from Knossos were evident everywhere; the young women of Crete had those eyes, that lustrous hair, the full lips. And who is to say that the olive groves of Mesara are not descendants of those olives that filled the amphorae with their oil, fuelled the beautiful pottery lamps, that kept the wheels of Minoan commerce running smoothly.

I was so young and earnest, walking the dry earth around the palaces with my notebook, trying to describe it all, trying to draw those elongated eyes, those goddesses, snakes in their fists, the priest-king with his headdress of lilies.

How much am I remembering, how much is dreaming? When I went with Agamemnon in his three-wheeled car to Kokkino Pirgos to pick up something for his mother, we stopped and walked away from the road. Did he carry me to the patch of myrtle or did I walk, alert for snakes? Knowing about them made every sound a danger. I do remember the smell of the myrtle as our bodies crushed the dark leaves under the sun, my back imprinted with a lattice of sticks. There were bees in the white blossoms. I do remember his eyes like almonds, his rough hands, and how I sat on a little terrace at Pirgos while his mother's friend asked questions in rapid Greek, sizing me up, then going into the house where there was the ceremony of water, a spoon, quince jam.

Sometimes he frightened me. He was strong, his arms thick with muscles, and he said — I think he said — "I want to make love to your bottom." I was unsure because of my imperfect Greek, his cursory English. But he was also funny, patient when I tried to learn a new phrase of Greek, and he was so graceful when he danced with the men who came to his *taverna* after several days at sea on a big boat. On such occasions, he'd spent the day cooking a special meal, and the *taverna* was closed to everyone else. I helped serve the dinner of lamb, cauliflower pie, zucchini blossoms stuffed with rice and dill. There were earthenware jugs of wine decanted from a barrel in the corner of the windowless cellar; I'd never tasted this particular wine — it hadn't ever been decanted in my presence — but one mouthful told me it was remarkable.

After dinner, the men all threw their plates to the floor. Maria and I cleaned up while they drank and musicians arrived to tune up. The lyra players were dressed in black, with bright sashes on their waists and across their heads; their instruments were shapely as pears. Some had hawk-bells on their bows, an ancient rhythmic accompaniment. The dancing was beautiful and wild, the drunker of the celebrants stepping onto the tables and turning, stamping, a few of them falling while the others shouted and clapped. When the musicians took a break, they held their lyras in their arms like beautiful women, stroking the wood with knowing fingers.

Did I love him? From this great distance, I don't think so. But it was exciting to walk with him and to listen to his heart when we lay down in the myrtle. I think of the way Eros involved himself in a game with a young woman, holding his hands out as though to suggest a path into the future while her chiton fell from her, her breast so young and exposed, more than two thousand years ago. I can believe how easily this happened. The boy who commanded her to take him on her back was unknown to her, a winged divinity. How quickly our childhoods recede so that we find ourselves recumbent in wild plants with an almond-eyed man or, braced against the earth, lifting him to heaven.

In those years, I did things I never intended to do. Some mornings I'd wake from a night I couldn't remember, head aching from too much wine. My body hurt and I wasn't sure why. Taking a towel, I'd walk down through sleeping streets to the sea and plunge in, swimming out in the buoyant waves until my arms were sore, turning to look back at a village impossibly beautiful and other in its secrecy. Maria would hold my waist in her hands and measure my hips, telling her son I was made to carry children in my body. I had enough Greek to know what she was saying, especially when she spoke slowly, almost lovingly. Swim farther, I'd tell myself, swim and swim until you reach Africa, then step onto the sand and begin again.

Two children at play. *The one who fails to overturn the stone carries the other, having his eyes blindfolded by the rider, until, if he does not go astray, he reaches the stone, which is called a dioros.* Only now the carrier is a girl. Her burden is a boy. She runs. It's a field of sunlight, ripe grasses under her feet. She tries to reach the *dioros*. She's laughing. And what's that, what is that lightness she feels at her cheek as the boy hoists himself higher? Her shoulders are aching but she runs. This is a game. She runs and her chiton drops from her chest; helped a little, who knows. What's that lightness? A wing. *A wing?* She is carrying the god on her shoulder and suddenly they are alone in a field of myrtle, olives, the dry pods of poppies, pale cyclamen, bitter herbs. Going astray with Eros. It has happened without anyone noticing.

I am writing now from a distance of nearly thirty-five years. A couple of summers ago, I met Joni Mitchell briefly in a local restaurant. We talked a little about Crete. "Where did you live?" she asked. "Not far from you in Matala," I replied, "in Agia Galini." Then, "Are you glad you went?" she wondered. And I knew she meant: are you glad you were brave enough, foolish enough — because this was before email and cellphones, when a regular telephone call was too expensive to contemplate, when letters took weeks to arrive at my Poste Restante address — a cardboard box in the Galini post office where once a day I'd search through the letters to see if any had come for me. I felt I was on the very edge of the known world.

"Were we ever that young?" I asked her. We both laughed.

I went to Matala once, with Agamemnon. At least three cafés called themselves the Mermaid, but he assured me that he would take me to the original one and he'd buy me a bottle of wine. There were caves where people lived — not Greeks but a ragtag bunch of Swedes, Germans, a few Americans, some Britons. A woman joined us at our table, and we offered her a glass of wine. She gulped it down appreciatively, then told us that she and her boyfriend were in the process of moving to a new cave. The ones on the upper level were choice, she said, and they'd been there long enough to be able to claim one that others were leaving. She was from San Francisco. Her children were out on the square, selling bracelets they'd made from beads and shells. A thin dog sat with them. They waved to their mother and went back to drawing pictures in the dirt with their toes. I think now of another *ephedrismos* figure, the joy of the two girls at play, and hope that those children in Matala were as happy.

The market sold shepherd's bags, green peppers, thick sweaters, used books, bright rugs, shawls, fruit, twine belts, and surprisingly (this was 1976) body jewellery of every style and description. And yes, when I looked carefully, most of the young people going to and from their caves had pierced noses, eyebrows, and lips.

"How long will you stay?" I asked the woman from San Francisco. "Forever," she said, quickly finishing her second glass of wine. "We can't afford to leave."

When I was passing the Acrocorinth on that beautiful morning, archaeologists were still excavating the Sanctuary of Demeter and Kore on its north slope. The excavations had been ongoing for decades, revealing more than a millennium of use, the sanctuary's origins in Archaic times transformed by Roman and then Christian use but intact at its core. This suggests a numinous quality to the place itself. That morning, if I'd looked up, I might have seen fieldwork involving various buildings associated with the Sanctuary — banquet rooms, a temple, a theatral area cut into rock, a room with a lustral basin for ritual cleansing. Excavation reports I've read since show a site with a history of intense Demeter cult activity involving votive offerings of cereals and other foods on terracotta trays, as well as sacrifices. Young initiates to the cult left figures at the site — thousands of these have been found by archaeologists, both intact and broken.

Demeter is generally accepted to be the goddess of grain, of fecundity, as well as the deity who presides over the mystery cults, foremost being the Eleusinian mysteries ("mysteries" came from *mystes*, an ancient Greek word for initiate). The rituals observed and performed at Eleusis sprang from Demeter's loss and recovery of her beloved daughter Persephone (also known as Kore) who had been abducted and taken by the god of the underworld. Inconsolable, Demeter neglected her responsibilities to ensure the success of crops. Eventually, she was reconciled to the return of her daughter for half the year and the Eleusinian mysteries celebrated this return in the form of elaborate springtime rituals. Feminine fertility, the earth's fertility: these were entwined, each mirroring the other.

This is the story of young women — not exclusive to that time and place — taken or lured from their mothers (often gladly, willingly) by the masculine force, no longer protected from their own incipient sexuality. I was fascinated to learn that some of the votives found at the Sanctuary of Demeter and Kore at Acrocorinth were *ephedrismos* figures. It seems such a natural thing for an initiate to bring to the goddess: emblems of childhood, some of them alluding to the moment when a young girl might find Eros on her shoulder, rather than the harmless boy with whom she thought she was playing. I recognize that moment in myself—a late bloomer, sitting by the sea with my book. Two men approaching me, an old man and his son, and I swear I never saw them coming. Wind filled with the fragrance of orange blossom and basil, lamb roasting in a nearby *taverna*...When Agamemnon took my hand, I knew what was inevitable. *If the olive groves didn't exist / I would have invented them.*

> Return flows calmly
> Forward and you follow
> Feigning indifference but pulling
> The rope to a deserted myrtle cove
> Not missing an olive tree
> Oh sea
> You wake and everything renews!
> — Odysseas Elytis, *Eros, Eros, Eros: Selected and Last Poems*

I have a bottle of Cretan olive oil that tastes of those months so long ago. Drizzled over tomatoes and white cheese, it has the power to transport me for a moment to those trees, bathed in sea air, nets spread under them to catch their bitter fruit. Angela is there, Yianni nearby, filling the donkey's panniers. Beautiful Eleni dreams of Demitreos while we finish our lunch.

Transported back to the grove where I lay with Agamemnon, a girl not ready for children herself, though eager for his body on mine, his calloused hands lingering a little too long on my buttocks. I renewed myself over and over in the clear sea and once saw a dolphin swimming so close I could touch it. Could have followed in its wake.

The young woman with Eros on her shoulder looks up to see the god, her cheek against his soft belly. I imagine her warm breath, the anticipation in her throat as he points to where they're going. On that bus from Herakleion, I watched the hills and villages of the island from the window, "little churches grazing / grass before the air,"[6] yearning to know their secret histories. I drew the goddesses, hoped for their strength, and all these years later, I still recall the smell of myrtle and the sight of olive trees rooted so deeply that it was unimaginable they would ever fall. If I close my eyes, I can still hear their leaves rustling.

Thuja plicata
Nest Boxes

> We must first look for simplicity in houses with many rooms.
> — Gaston Bachelard, *The Poetics of Space*

I grew up with cedars. To some degree they defined the way I apprehended space and time. The ones I remember best were at Goldstream Provincial Park. Huge and shaggy, they grew near the river where I went to look at spawning salmon each fall and returned in spring to look at the fry darting through the clear water. A trail meandered through the cedars to a salt marsh and the estuary of the Goldstream River. The park was densely green. In spring, the smell of black cottonwood leaves unfurling was heady, their resins scenting the entire area. Moss-hung bigleaf maples and their honeyed blossoms were alive with warblers — orange-crowned; yellow-rumped; black-throated grey; Townsend's; MacGillivray's; and Wilson's. I remember it was a good place to see trilliums and the beautiful shooting stars with their swept-back magenta petals. There were also skunk cabbages in the low damp areas.

Occasionally, I saw bears. In fall they feasted on the salmon, and in early spring, newly awake, they ate bright green leaves, their scats glowing with chlorophyll.

The cedar roots ran along the trails like mountains on a relief map, emphasizing the verticality of the landscape. The trees themselves, or at least the ancient ones the park was known for, were heavily buttressed at the base. Their trunks were fluted and ridged, the bark coming away in places. It was not difficult to imagine oneself in a cathedral, one hung over with a ceiling of blue or cloudy grey, punctuated with birds. In the high canopy, waxwings and evening grosbeaks fed on seed cones and insects; fall and winter, scores of bald eagles feasted on salmon and surveyed the world from the tallest cedars.

Some of the cedars were more than five hundred years old. Older than the city I lived in, older than my country in the name given it by latecomers. In the infancy of these trees, explorers measured the altitude of the sun and other celestial bodies with their astrolabes as the oceans carried them to North America; John Dowland's *First Book of Songs and Ayres,* first published in 1597, was the most often reprinted music book of its time;[1] and First Nations people on the islands off the western continent had been building houses of their broad planks — stitched at the corners with their plaited branches — for thousands of years.

When I was growing up, my family moved every two years. My father was a radar technician in the navy, and he would be transferred from Victoria to Halifax, from Halifax to Victoria, from Victoria to the radar base on Matsqui Prairie, back to Victoria. We never owned a house. We'd stay in motels for the first part of most transfers, having outgrown the family housing offered by the navy; my parents drove to possible rental houses with my three brothers and me in the back of the station wagon and our black Labrador, Star, in the very back, drooling as she hung over the seat.

Moving was exciting, and also a little sad as I thought of all I would miss—the fields behind our home in Matsqui; fishing with string and bent pins at Herring Cove near Halifax; and the cemetery in Fairfield where I squeezed into crypts and communed with the dead. For weeks, my mother made lists and tried to organize what we owned. My brothers and I chose favourite things to take with us on the journey—a book; a stuffed animal; baseball gloves for games of catch in campsites; binoculars. Then a moving truck would pull up in our driveway and teams of men packed up our belongings, wrapping breakables in creamy paper and fitting them into large wooden tea chests, wrapping padded blankets around the furniture, then loading everything into the truck. The house echoed with the loss of our possessions and my mother did a last-minute sweeping of the floors, polished the windows with newspapers and vinegar.

A truck eventually pulled up in the driveway of the new house and everything was unloaded. My mother cried to discover that cherished plates had been broken or a lampshade crushed. The furniture was arranged in the rooms and I'd lie in my bed at night and try to orient myself by remembering my old room. Closing my eyes, I pointed my finger in the darkness to the window. Waking, I was surprised for weeks by the unfamiliar light.

There was always a moment I waited for, the moment when my mother replied, "Yes, I think so," to the question I posed daily after one of these moves: "Are we settled yet?" Settled meant that we knew where things were—light switches, the spaghetti pot, a hammer to bang in nails for our pictures, our winter jackets. New friends knew where to find us. Letters arrived in our mailbox.

The last family move was in 1969, when I was fourteen. My father retired from the navy and we moved from Matsqui to Victoria, where a job waited for him at the dockyard in Esquimalt. A house had been purchased, the first and only house my parents owned. The sale had been accomplished on a weekend trip to Victoria a month or so before we moved. There were a few requirements—enough bedrooms, a paddock for my horse (in Matsqui we had rented a house on a farm and my lifelong wish for a horse had been fulfilled), close to schools. There were also a few hopes—my mother wanted a dining room, a fireplace, and two bathrooms.

I have fond memories of a house we lived in when I was in grades one and two, a house with a pagoda roof and an attic room accessible by ladder, doors that opened with crystal knobs, a bark-burning stove in the kitchen, a greying cedar trellis in the leafy backyard, a small neighbourhood park right across the road; I imagined that such elements might be a part of the new house. None of these were to come true.

The house we moved to was an ordinary 1950s bungalow. There were three bedrooms, a small bathroom, a small kitchen, an adequate living room. But there was also a basement, and a plan to rough in some rooms down there behind the furnace where small windows, non-opening, gazed out to the carport. There wasn't a paddock, but the house stood on nearly an acre, the back part of it wild, so we would fence an area for my horse; he remained behind in Matsqui at the farm of a friend until we were ready for him. A few junipers in the front yard, a hawthorn, some pines in the backyard. A vegetable patch which my father would annually rototill and whack with a shovel, swearing at the clumps of hard clay that refused to crumble.

We moved to that house well after the school term had begun. It seems my father had gone through a kind of crisis, half-wanting to buy into a sporting goods shop in Abbotsford, where he could have worked as a gunsmith—his hobby—but knowing also that a job waited for him in Victoria with known income, benefits, the things he was accustomed to and upon which his family depended. My parents argued in the night those last few months in Matsqui and finally we moved to Victoria to the house which didn't fulfil anyone's dreams, and where a patch of overgrown sour grass waited to be fenced for the arrival of my horse. We stayed in the Cherry Bank Motel while we waited for the moving truck to arrive, and we were registered in schools in the area, our father driving us each day from the motel. It was painful to be the gawky girl introduced to the class a month into the term.

All my life, I have wondered at the feeling I have in particular houses, usually ones in which no one lives any longer. I've felt it in Point Ellice House in Victoria, where members of the O'Reilly family lived for nearly a century and where the rooms are arranged in tribute to those days; felt it in abandoned farmhouses on Sumas Mountain when we'd come across them on blueberry picking expeditions and where a tattered remnant of wallpaper, neatly cut (if mouse-eaten) squares of newspaper on a nail in the outhouse; or a rusty cookstove spoke to me of the deep legacy of belonging and loss.

Once, in Utah, I wandered around a cabin in the Dinosaur National Monument Park and felt the presence of the family that had lived in that place so vividly that I had to wipe tears from my eyes. A tire swing hung from an old cottonwood, clematis covered the roof of the cabin and foamed over the windows in cascades of white blossom, and a few milk cans stood battered and empty outside the collapsing barn.

Sometimes a house seemed as though it was waiting for its family to return, furniture still in the rooms, a kettle on a stove. There was a low clapboard cottage in the woods near Elk Lake, where I rode my horse, and its windows seemed to me a study in patience, as did the lilacs which bloomed in spring, in anticipation.

I would think, Entire lives have been lived in these houses, and would be filled with something like sadness, but not quite. Later the word *nostalgia* settled into my lexicon with such ease that I knew I had been waiting all my life for it.

When I was a young woman, I travelled through Europe with a change of clothing in a knapsack, and imagined myself into a shepherd's hut on the south coast of Crete, my lover Agamemnon bending to enter its single room and showing me its hearth, a small opening in the roof to take away the smoke.

There was a room in the commune near Grasse, in France, where I was taken by friends for lunch. We were served food grown and prepared on the property—even a glass of the brandy made in the cellar, barrels scented with oranges from the trees providing shade to the terrace.

Later, I lived in a cottage on an island off the west coast of Ireland and planned to live there forever, finding in its stone walls and windows facing the north Atlantic a solace of long occupancy. People had inhabited the island since Cromwell's cry, To Connaught or to Hell! drove them there in a desperation of survival. My cottage wasn't from those days, but the ruined and tumbled buildings on the island probably were. I loved the wind, the hedges of fuchsia, and the jugs containing milk the colour of primroses that my landlord brought me.

So many living things require and actively seek a home. All around me as I write, birds are making nests. Some are choosing among boxes we've nailed to trees. Bears have awoken from their winter dens and are walking the high trails, eating new grass and insects — the mountain itself their home. On the shore, hermit crabs awkwardly drag their whelk shells.

I remember watching a muskrat on one of the creeks on Matsqui Prairie build a dome-shaped house of cattails in a quiet eddy; from my place under trees on the banks of the creek, I imagined myself inside that place, surrounded by the smell of water and whiskered fish. Even octopi have their dens, the larger ones choosing underwater caves and the smaller ones, old bottles and moon snail shells.

Home offers protection and exerts a strong influence on organisms, shaping them physically as well as spiritually. Imagine knowing a place only briefly — a few months — but then remembering it so precisely after a period of three or five years, that it's possible to swim a thousand miles to find the tiny stream where conception and birth took place. I think of this every fall and winter when I watch the return of the Coho and pink salmon in local streams.

"Home," my 1974 *Concise Oxford Dictionary of Current English* tells me, means, "dwelling-place; fixed residence of family or household." When I was in grade seven, my teacher asked the class to write a composition on what home meant to us. I still remember the rich feeling of putting into words all I wanted, all I hoped for. My piece wasn't about an actual place — that year was the second or third we'd spent in a patchy old house on Harriet Road. The house itself and the circumstances did not characterize home. My room was in the basement:

I'd been caught shoplifting at a local grocery and spent long hours feeling remorseful in my underground quarters next door to my oldest brother, who made my mother cry because he kept hiding pictures of naked women and she thought he was headed for hell (a lonely boy I could hear in the night talking to an iguana he kept for a pet). I came back from school one day with scabies caught from Andrew Elliot who sat next to me and who never washed. Every night for some weeks I had to stand naked in the bathroom while my mother painted the open sores with some sort of disinfectant. In my essay, I wrote about ideals: the warmth of a fire, the taste of hot chocolate on a cold winter day, the fact that there was always enough hot water for a bath, and enough dumplings in the stew for each person to have two. My teacher read my composition to the class and I listened as though to something written by someone else, or if not exactly someone else, not really me either but a finer self. It was a moment when I knew that words could do something other than fill in space.

A kiss led me to the truest home I'll ever have. After several years of travelling, I was paused in Victoria, waiting for something to happen. I expected to return to Ireland, where I'd been happy, once I'd earned enough money. To that end, I was working in a bookstore in the mornings. In the afternoons I'd cycle back to the tiny apartment on Fort Street where I was trying to find my writing voice at a desk under the window looking out onto Garry oaks. Then I met a poet in Victoria — he'd come from Vancouver to give a reading at the Open Space Arts Centre — and he was walking me home from a party, through the dark streets of Rockland. Stopping in front of the Art Gallery, he kissed me, a moment that began the rest of my life.

At the age of twenty-six, I helped the poet — now my husband — erect an old blue tent on a plywood platform on an Easter weekend. Our son, two weeks old, waited in his car seat to be moved into the tent wrapped in a swaddle of blankets. In the back of the car, among the bags of diapers and the week's provisions, was a ball of twine and a plumb bob. With these, we intended to mark out the shape of a house on a bluff, facing southwest. We ordered piles of lumber (a sling of north species 2x4s. We culled the cedar out for decks.) and then arranged them into the makings of a house. Each pile contained within it the dimensions a heart tells to the hands—to the saw, to the hammer and nails, the rooms accumulating until there were enough for a home.

Spirit Level, Plumb Bob

We began to build a house with only a hammer and a few chisels. Maybe a multi-headed screwdriver—I can't remember; it might have come later. Of course we bought tools as soon as we knew we needed them. A Black and Decker 5¼-inch circular saw. An Estwing hammer. A line level, a carpenter's level, and a brass plumb bob. There was so much measuring and levelling that it's all a blur now, though I remember how hard we both worked, falling into our bed at night, exhausted, muscles we didn't even know we had strained and aching.

We'd come to our land for three or four days a week at first, loading up tools and food in our car, along with our newborn baby boy, Forrest, and everything we needed for him—diapers, clothes, blankets.

The first night, we camped with him in the tent we'd set up on a platform of plywood with tarps over it that were tied to small cedars on each side. The tent was cosy but cramped. Everything had to be kept from the sides so rain wouldn't seep in. Forrest was the only one who slept. We were worried he'd be cold or, well, we didn't know what, exactly. We hadn't been parents for long. We had a foamie for our bed, with sheets and a down sleeping bag for a comforter, and it was warm. But it was also April, so I remember it rained more often than not. I'd lie awake, waiting for the baby to cry. John lay awake, waiting for the tent walls to let in water. Forrest slept between us, his head warm in a little knitted toque. When it began to get light, loons warbled down on Sakinaw Lake and once something screamed nearby, uncannily like a baby, and our large English sheepdog cross, who was sleeping under the tarp on her rug, struggled to get under the tent platform. Later, I realized it must have been a cougar.

First we built an outhouse. This was a requirement of the Regional District building code; and in fact, we realized that if we could build four walls with a shed roof over them, if we could hang a door with the obligatory quarter-moon screened for ventilation, then we could probably build anything. And what luxury, to sit on a toilet seat with literary magazines at hand, instead of crouching in the woods, the dog sniffing at our butts as we did so, and then discreetly burning used paper in the fire.

We scraped our building site clean of salal and Oregon grape and measured. Then we made batter boards — each corner of the house site framed with two horizontal boards at ninety-degree angles, attached to stakes, and perfectly level. We used the carpenter's level for this, setting it on the boards until the small bubble in the glass vial holding ethanol balanced in the centre of the tube, telling us we had horizontal level. When Forrest cried, I'd run to the tent to nurse him, wrapping us both in a blanket as I pulled my shirt aside.

After the batter boards, we dug holes for the footings. John rented a rock drill and drilled into rock for the footings which needed to be anchored with rebar. Some did, some didn't; it depended on how far down we needed to dig, where grade was, where rock was. Wooden forms were built for several footings. When we didn't need to drill, the rebar was sunk into the concrete, which was made in a wheelbarrow and shovelled into Sonotube or a wooden form. The plumb bob was used to find the centre of the footings, a vertical level.

We made our meals at a table we'd built out of tongue-and-groove boards nailed to a frame of small cedar logs. We had an old Coleman stove, and a large enamelled pan — for a sink, a salad bowl, a bathtub for Forrest. We also made a campfire in a ring of stones and kept a pot of coffee warm on the stones; we cooked meat on a wire grill over the fire. We could also boil a kettle on the fire if we weren't in a hurry for the water. The water came from Ruby Lake. We'd take down our big twenty-litre container and dip it into the deeper water. Sometimes a little gravel ended up in our mugs.

Every step was wondrous. The cedar posts on the footings, then the long beams of strong fir. At that point, our building site looked like sculpture, silent prehistoric animals waiting on the bluff. We built the walls on the platform created by nailing joists to the beams on sixteen-inch centres, then setting plywood on top, using chalk line to tell us where to nail. The dark red chalk stained my hands, the mark of a builder. We lifted the walls ourselves, apart from one or two very heavy ones. And then we'd ask a neighbour or a friend to help. John would carefully brace the walls with 2x4s so that they couldn't fall over the side of the platform, but each time there was anxiety as we lifted and held, one of us holding the level and suggesting adjustments, the other nailing down the bottom plate.

Every step—the sheer weariness of holding ceiling rafters in place (it took some time to figure out how to create a bird's mouth notch), strapping for the cedar shakes for the roof, framing a doorway, nailing down plywood. Understanding the role of lintels, those horizontal structural members supporting a load above a window or door.

How did we ever do it? How did two poets with a small baby build a house when they had no experience beyond building bookshelves out of pine? It never occurred to us to buy plans or consult an architect. We had the building code, and that told us what we needed in terms of requirements and standards. But I marvel at how John could envision our house from his drawings. We had agreed on sizes for rooms and their placement, but I was no help at all with the actual plans; he drew them and got them blueprinted and then approved by the Regional District's building inspector.

Will I like what it will look like? I'd ask, trying to imagine a kitchen from the plans, how the windows might let in light. Where will the sink be? Where will our bed be? The drawings showed the dimensions, elevations, each room's relative size, to scale.

> For, in point of fact, a house is first and foremost a geometrical object, one which we are tempted to analyze rationally. Its prime reality is visible and tangible, made of well hewn solids and well fitted framework. It is dominated by straight lines, the plumb line having marked it with its discipline and balance.[2]

I had no spatial sense at all but nailed and lifted with blind trust, unable to translate structural materials to interior space. Drawings spoke of rectangles, clean and elegant. But will there be a windowsill for a plant? Where will we sit to watch sunsets?

We built four nest boxes in the year leading up to my fiftieth birthday—nearly half a lifetime away from that kiss. Three were for us and one was for friends.

I found the plans for the nest boxes in a gardening magazine. The plans were simple—little houses constructed of rough cedar, a clever arrangement for opening the top (a sloping length of cedar board which would repel water like a shed roof), and there were several dimensions given, depending on the birds one wanted to attract. Some birds like an oval entry; some prefer a round one. Some like a perch, whereas a perch can also be a means for larger birds to rob the contents of a nest. Some birds, like purple martins, will live in multi-family constructions but others, like swallows and finches, like privacy.

We were hoping for violet green swallows. Fifteen years ago, elderly friends gave our young children a nest box which we nailed to the top of a post holding some of the wire which surrounded my vegetable garden. For several years, swallows nested in that box. We'd see the pair swoop in come April, excitedly exploring our house and garden, then entering the painted box with little squeaks and chirps. They'd disappear, only to return a few weeks later. The male would sit on the wire that conducted electricity into our house, while the female carried nesting materials through the opening. Then the male would bring in a few bits and pieces and spend time examining and adjusting the nest while the female sat on the wire. They'd take a break from this pattern for a few minutes of ecstatic flight, their wingtips touching in the air, their ardour breathtaking to those watching from the ground. I thought of an aria sung by Magda in Puccini's opera *La Rondine*, its notes echoing the beauty of the swallows' flight, their courtship, their residency in the shelter of our garden. When the swallows first appeared each year, I'd play the opera as an homage — Montserrat Caballé, recounting her dream of a revelatory kiss.

I don't know very much about the mating habits of swallows, although I understand they are monogamous. Our pair seemed quite affectionate with each other. When the young began to peep in the box, the parents were very solicitous, removing faecal material, bringing endless supplies of insects to open beaks through the opening. A little more than three weeks after we first heard the peeping of babies, the young fledged. The family still remained together, the parents teaching flight manoeuvres, the young practising over our garden, the entire family feeding on swarms of insects. Then, one day, they'd be gone.

The original box eventually began to fall apart, about thirteen years after our friends brought it to our children and several years after they'd died. The roof split apart at the top, and the bottom began to rot. Two winters in a row, I removed it from its pole and cleaned out the mess inside, drying it and fitting roofing felt over the cedar shakes, hoping it would last one more season. In the meantime, I put another box up in another location, but no birds went near it. Perhaps the opening was too big or the wrong shape.

And the time came when the swallows rejected the original box, too. I sought out plans for nest boxes specific to swallows because I missed their presence — the high, tremulous swoop as they courted, the eager noise as they chose their seasonal home, the chorus of infant birds asking for insects. It was as much a part of spring as the first rhubarb or Apeldoorn tulips opening their golden bowls to the sun.

We moved into our unfinished house on the eve of John's thirty-fifth birthday: December 1982. The walls had been finished with plaster and painted, the windows were in place but had no trim or sills, the exterior doors were hung, of course, but there were not yet interior doors, apart from the bathroom. We had a long trestle in the kitchen with a makeshift sink, though a new stove and fridge gleamed, plugged into the electrical outlets that John and my father had laboriously wired into place, long strands taking power to all the rooms, a chart detailing their journey from the panel on the wall by the fridge.

My brother helped us move. He and John rented a truck and filled it with our bits and pieces of furniture from the house we had leased in North Vancouver. I went on ahead in the car with Forrest. I wasn't much help with the lifting because I was heading into the final month of the pregnancy that resulted in Brendan. The plan was that I would take an early ferry and have time to prepare a hearty dinner — steak, baked potatoes, salad, and crusty bread — for John and Gordon to enjoy once they arrived with the truck.

We hadn't counted on a windstorm. My ferry sailed on time, but they were delayed in Horseshoe Bay because the ferry's generator was supplying power to the village, which had lost its power due to fallen trees on power lines. When I arrived at the house, I discovered that there was no power there, either — and the large picture window in the living room was leaking water around its edges.

Several people had commented that our house was the first they'd seen with the cedar shiplap siding applied horizontally. Most people used either bevelled siding or else they nailed the shiplap on vertically so that any water collecting in the channels would run to the ground rather than sit in the grooves and perhaps seep behind into the building paper. Was our insistence on doing it our own way proof of our naivety? Folly? The wind was blowing hard. I made a fire in the woodstove, though smoke kept blowing back into the kitchen, and I lit the oil lamp — we only had one in those days. At least the phone was still connected, and after hearing over the battery-operated radio that ferry sailings from Horseshoe Bay were delayed, I waited for it to ring.

I remember how bleak it felt, sitting by the fire in an unheated house, a single oil lamp providing limited light, knowing that this would be my life — this house on this bluff facing into the wind. I waited for the sound of the truck. Many hours later, on the dark driveway, I stood in front of the sliding doors (which opened into space; the deck came later), holding the lamp and hoping they could determine the reason everything was black was that the power was out. I expected them to be hungry and tired, but knew I couldn't bake potatoes or do justice to a steak on the Coleman stove I'd brought in from outside. Forrest, twenty-one months old, was asleep in his temporary crib.

Gordon and John arrived cheerful and full of a dinner they'd treated themselves to in a Horseshoe Bay restaurant with a generator of its own. They brought laughter into the darkness, immediately opening wine and regaling me with stories of negotiating the winding highway up the coast, past fallen trees and branches rushing by in the wind. They were happy to eat bread and cheese by the fire, filling their wine glasses over and over again. But in our improvised bed that night, with my brother sleeping in a room nearby, John and I talked quietly about the storm, the leaking window, and how we might have made a terrible mistake — and not just with the horizontal shiplap siding. Holding onto each other in the dark as the wind battered the house, we wondered if maybe we should have bought something in the city, harnessing ourselves to a mortgage and the necessity of two incomes for the rest of our lives. Everything seemed gloomy and we were very far away from what we'd known and loved.

The next morning dawned brilliant and calm. The wind had died, the power was back on. Gordon and John got themselves organized to unload the truck and arrange our furniture in the bare rooms. First, we ceremoniously laid our wool carpet over the bare subfloor in the living room, where it brightened the plywood and caught the light streaming in the picture window. In the clear day, John could see that the water coming in that one place by the big window wasn't because of the application of the siding, but because he'd hadn't caulked that particular place adequately to seal the window flange. This was easily remedied.

We had our first Christmas in our new home, with my parents and my brother as our guests. We had a big fir tree in the entrance hall. Who needed kitchen counters to make a feast of roast turkey and all the traditional accompaniments, including John's famous sherry trifle? By the time Brendan was born in late January, we'd had a friend build kitchen cabinets out of yellow cedar. We bought a sale lot of terracotta tiles for floors and counters — and the tiling was done in the summer, when I could take our young sons away for two weeks and let John tile day and night without distraction. There were doors for all the rooms.

"Sometimes the house grows and spreads," wrote Gaston Bachelard in *The Poetics of Space*, "so that in order to live in it, greater elasticity of daydreaming, a daydream that is less clearly outlined, are needed."[3] What wasn't included in our plans, so carefully drawn by John by the light of small reading lamp at his desk, drafting ruler at hand and a selection of sharpened pencils, was the eventual arrival of a third child. There was one bedroom for our sons to share, and a study for John and me, which also contained a sofa bed for houseguests. The entire second storey, a twenty by twenty square foot space with plumbing roughed in for the day when we could afford time and materials to finish a small bathroom at the top of the stairs, was our bedroom. We had intended to divide the space into two rooms, but once it was framed, we loved its views and airiness, and left it open.

After Angelica's birth, we began to plan an addition. By pushing out the south wall of the boys' room, we then built two more rooms, reasoning that the boys could still share; we bought bunk beds for them. Once Angelica was old enough to need a room (she was sleeping with us while breastfeeding), she could move into the very small one between the larger one for the boys and their old room, which would become a playroom. The addition would have a flat roof that we could use as an upper deck, a small sunroom leading to it from our bedroom.

Sometimes the house grows and spreads: that small addition lasted for a few years, and then it was clear we could use more room. Personalities grew as rapidly as limbs. Out came the drawing paper, pencils, special ruler, and a plan to extend in another direction. By taking out part of the eastern wall of the playroom, we could add two more rooms — one with a small step up to accommodate the rise of rock beneath it. By knocking out the wall between the two earlier rooms, we could create a larger room there for one child, and then each of the others would have a room in the new addition. The flat roof on top extended the deck off our bedroom and it was also a good idea to build a cosy study for John. The playroom evolved into a library to hold the bulk of our family's book collection.

By now John had familiarity and skill with his tools. He knew how to make the best use of materials and how to set priorities, rather than daydreaming of windowsills and sunsets, the way I did.

We decided to have a few of the cedars on our property cut down. They were on the northeast side of the house — small trees when we'd first built in the early 1980s, but now towering and full-branched, and too close for comfort during intense winter storms. Gradually, too, their fallen fronds soured the soil where I was trying to grow roses and there was too much shade for anything else to thrive.

It always feels a little wrong to cut down a healthy tree. We thought about it and talked about it. On the one hand, on the other. And then we called in a team of guys. They had no qualms about taking down cedars. "Weeds, they're weeds in this climate," one of them said as he prepared his saw.

I tried not to be home on the day the cedars came down, but inevitably saw part of their demise. Even though the tree fallers went up and limbed the big trunks before cutting each in segments, there was a moment when one section — I'd come home expecting everything to be done, but the team had arrived late — hit the ground with a big *whoomph*. It was the biggest tree. The log that came down so hard was a good size, and we arranged to have a portable mill and sawyer come to cut it into rough boards. We hoped to get a 16-foot length of 4x6 out of the big trunk to replace a beam across our patio. A wisteria, nearly twenty-five years old, clambered across the beam from the woodshed end, creating a green bower, and at the other end, a New Dawn rose spilled its soft pink flowers over the rough wood. The stump of the biggest cedar measured more than a metre across. The guys cut it level, using their huge saws as skilfully as cabinetmakers, so I could put a large planter on top.

The mill arrived, pulled by a pickup truck held together with wire. The sawyer had been recommended by several sources — but always with the proviso not to get downwind of him. He had just been in hospital to have a steel rod inserted alongside his spine (can this really be possible?) so he'd reluctantly brought a helper, a staggering fellow missing several fingers. His job was to carry the enormous lengths of cedar log to the mill where they would be sliced into boards.

The smell that day was not, as feared, of the sawyer's odour (though it could easily be detected when I passed him coffee on their morning break — something extraordinary, like animal fat and mysterious unwashed corners of the body and clothing steeped in both wood and tobacco smoke), but of the spicy scent of fresh-sawn cedar. The boards were beautiful as they emerged from the end of the mill — pink and salmon, the grain an intricate story of age and weather. Several times I was horrified to see the sawyer lifting logs himself and imagined an emergency, the ambulance negotiating our rough driveway, paramedics removing his ripe body from the ground with a metal spike exposed at the back of his neck.

The pile of lumber grew — the beam, some 2x10s, 2x8s (these were full dimensions, as the boards were unplaned), some planks which began as one dimension but then tapered as the logs narrowed. I could see them as benches or tables, balanced on stumps. I kept touching them. Their surfaces were damp, the inner mysteries of the wood released to light. On one chunk of wood, hardly a board, the grain formed an eye, elongated and ovoid — a god or a raven staring out. When I smelled my hands afterwards, the incense lingered, familiar and sibylline.

I was inside, doing some task in the kitchen, when John came in with his fist closed over something. What, though? Too early in the year for tree frogs, too early for an unexpected gift of raspberries brought dewy from the garden on a July morning, their tang on the tongue a promise and memory of every summer.

He opened his hand. Five seeds. His eyes shone. "They were inside an area of rot I was prodding at with a screwdriver in one of the big planks."

"Show me."

So we went outside to look at the board and its open hole, where the rot had been crumbled out with the screwdriver, a few cubes of diseased wood on the ground beside it. By a little bit of deduction, we realized the board had come from the inner section of the lower trunk. When we went to the remaining stump, we saw the corresponding section of rot right at the tree's old heart. It looked like something called "brown cubical rot," which forms a seam up the centre of the butt. It's introduced by the mycelia of a fungus that grows on the trunk. When we first came here in 1980, the tree was young and stringy. I don't remember a wound where rot could have begun, or a shelf of bright fungus — but given that we were building a house and raising three small children, there was a lot I didn't notice in the course of my days. I do remember that, in later years, there was a cicatrice low on the trunk where the bark had healed. The tree was popular with squirrels.

The seeds were obviously squash of some sort. Creamy, a little sticky with resin, a strange gift to show up in a clump of rotten wood at the heart of a tree. We spent time over coffee reconstructing the narrative of those seeds, remembering back to a particular summer, twenty-five years ago, when I'd staked out a vegetable plot, 25 feet by 25 feet, and tried to improve the rocky soil by digging in seaweed and anything else I could get my hands on: a bucket of chicken manure from neighbours, mulch from under the bigleaf maples, sandy run-off at the bottom of our steep driveway. Our two small sons played in the dirt that eventually was raked and seeded to what passes for a lawn, enhanced with wild moss. I planted pumpkins that summer, wanting the beauty of their orange globes to remind me of harvests. The plants had spread out with wild abandon, and a few of the pumpkins forgotten under salal beyond the boundary of the vegetable garden. What an opportunity for squirrels. And a few seeds tucked into a likely crevice, fresh and raw, in the trunk of a handy tree to provide a winter meal were forgotten, maybe after the tasty fungus had already been knocked off and eaten, forgotten as the tree healed around the small rent.

I planted the seeds, and three germinated within a week. I transferred them to the vegetable garden once it was warm enough. I'd love to say that they thrived and produced a huge crop of pumpkins (for they were in fact pumpkin seeds), a testament to my green thumb and the seeds' inherent fertility. The truth is, they didn't do much of anything. They grew a little, sent out tendrils to hold fast to the stems of kale. A few blossoms, a few tiny green pumpkins which never matured.

I was disappointed — but too busy with jam-making and canning to linger too long on this failure. I decided that the true magic was in the finding. That hidden in the heart of a tree was unexpected treasure, a mnemonic to take us back to our beginning days on this property, when our garden grew beyond us, when we carved the thick skin of pumpkins into faces on the night of All Hallows, lit from within by a short length of candle, to stay off the spirits that crossed the boundary between the living and the dead.

How the time passes quickly so that a sapling—I just looked out to see it—leaves a trunk almost a metre across when felled, its years, the weather contained in a narrative of rings. A seed waited for twenty-five years inside that tree to have its chance to become a pumpkin, however small and green the result, and the children who crouched under the limbs to while away a hot summer day have become scholars and lovers, their lives elsewhere except for a few days a year when they walk the old paths, sit by the fire that continues to draw us to it each morning, a fire started with split shakes of the original roof, now silver with age. How time passes, how everything we knew is stored in our own bodies—the dull ache of sleepless nights, the sharp yearning for love, the sorrow of these empty rooms once filled with children laughing, fighting; their books, their toys, their filthy socks, and tiny overalls. One boy still sits under the original nest box (though I know it's not possible, he lives in Ottawa) with his notebook, trying to sketch the swallow nestling that hangs out the opening, saying, Don't fall out, Parva! Be careful. And I stand out among the trees, under stars, while the moon thins and fattens, turns soft gold in autumn, hangs from the night's velvet in February, draws me out on summer evenings to drink a glass of wine while owls fill the darkness with that question: *Who cooks for you, who cooks for you-all?* It was always me and I never once minded.

I've been watching robins this year. One pair built its nest on the downspout of our print shop, a short distance from the house. From the kitchen or the porch, we could see the progress of the nest and then the familiar sight of the female perched on it. There were robins in the same place last year and that couple raised two broods. I love peering out, with binoculars for the best view, to see the patient bird incubating her eggs, rising to perch on the side of the nest to turn the eggs, then taking a short break to find a meal for herself while her mate stays near to protect the nest. It takes about two weeks for the eggs to hatch, and then the mother robin never seems to rest, darting out and back to bring worms and insects to the increasingly active brood.

Sometimes all I can see are three beaks open to the air. And then three gangly young birds carousing in the small space and calling for more food. It only takes two weeks for them to grow to adolescence and leave the nest, each perched on its woven precipice and then soaring out into the world.

Once we were lucky enough to see the last of the clutch leave, a sweet moment as the bird leaned forward eagerly while a whole gaggle of robins called and flapped from a nearby cedar. Finally it just...flew. Imagine just knowing how! Just pushing off from the nest and flying, something many of us dream of doing. I've read that the male robins continue to feed the offspring for two weeks after they've left the nest and then they're on their own. Depending on the time of the season, the female will be nesting again, prepared for the hours of waiting for her eggs to hatch; then willing to feed the rapidly growing chicks for the two weeks it takes them to mature.

This year, the downspout couple raised one family and then either they disappeared or else they are the same birds who built on the other side of the house, on an elbow of wisteria just outside my study window. I watched this nest from my desk, looking up from my work as I'd hear a rustle—the mother returning with food for the three young. After the babies finally left, the mother spent some time rejuvenating the nest; she brought fresh moss, fresh grass; and I thought how wise she was to have chosen the site in the first place. The wisteria leaves make a shady canopy over the southwest facing nest. But she didn't stay, perhaps deterred by John who was building new steps for a reconstructed sundeck nearby. (He'd put off this project until the young had flown.)

Reading about robin mortality rates, I was surprised to find out that only 25 percent of robins survive until early November of their first year. Life expectancy is two years. The hard work of the industrious parents, raising up to three clutches a season, is not well-rewarded. Yet robins seem ubiquitous. Driving along the highway in spring, one sees so many of them at the roadside, flying up in challenge as the car approaches. (This rash bravado might be the very thing that limits their survival rates, or at least for those 25 percent who survive past November. In spring and early summer, I often see dead robins on the side of the highway, though the ravens and vultures make short work of the carcasses.)

And there are predators. One night, before the wisteria family had flown, we were awakened by two barred owls very near the house. I know they are capable of taking robin eggs and chicks. For about two hours they chorused back and forth to each other, their eight-note call — *Who cooks for you, who cooks for you-all?* — with its drawn-out final quavery note becoming shorter, more urgent: four notes. And finally just a long descending throb, right by our bedroom window. I wondered if the parent owls were perhaps teaching their offspring to hunt, and if nesting birds near our house might be the prey.

So now it's back to the downspout and the mother is on that nest as I write. I loved watching her prepare the nest back in April. There had been one there in the past and I know that sometimes robins simply build on top of an old one but that earlier nest had fallen, a perfect construction of woven twigs and moss, held together with mud, and then lined with grass. The new nest took a few days to build and, at the end, the bird crouched in it and plumped out her body, turning as she did so. This formed a cup to the dimensions of her body. She carried wisps of grass to it and then I think she laid her eggs, one a day for three days.

This time around — it's early July — she simply reoccupied the nest that she had used in April, bringing a little fresh grass for her new family. If we get too near, she glides out and is back again before we know it. I love to hear her mate singing morning, noon, and night, the long rising and falling notes clear and bright.

Of course by now you will know that I am talking about my own family — three children raised in our homemade house, nurtured and loved, and coaxed easily from the nest with every hope for their long survival. Oh, and their return! "So there is also an *alas* in this song of tenderness. If we return to the old home as to a nest, it is because memories are dreams, because the home of other days has become a great image of lost intimacy."[4]

Think of those chicks crowded in that bowl of moss and mud, jostling and agitating for the food from their mother's beak. That first glide from the nest into thin air, the vast blue yonder, must've been heaven. Yet for days after, I see the mottled immature robins perched in the cedars near our house, uncertain about the future, perhaps, and reluctant to leave the actual palace on its elbow of wisteria or downspout.

This spring we cleaned out the nest boxes again, propping a ladder against their respective trees — an arbutus, a fir, and a small cedar cut down a few years ago, limbed, and set in place as a garden post. This last location was where we'd nailed the first box, the one that welcomed swallows and where Forrest called to Parva on summer days long ago. Each box contained remnants of a nest, a small cup of dried grass and moss and a certain amount of hair from our golden retriever.

At least one chestnut-backed chickadee couple nested in one of the boxes last year. We saw them checking it out, darting in and out excitedly; and then one of the pair sat on the clothesline while the other took in threads of moss or lichen plucked from branches of ocean spray.

Maybe the other nests were older. Maybe I never noticed. The years pass and the summers enter the rich tapestry of memory so that we ask, When did we plant the ornamental cherry tree? Or the fig tree, laden with green fruit as I write, or when did we swim by moonlight, or cook sausages in a grove of trees on White Pine Island among flowering yarrow and sweet golden grass? Which was the last year we all lived in this house, dogs eager for the children to run with them or take them up the mountain to enter the cool creeks in early morning while spiderwebs jewelled with dew hung across the water?

I am still hoping for the swallows to return, though it's too late this year. We saw them for a few brief days in spring, flying ecstatic over our roof and garden. And we know they nest in multitudes down by the lake, where a fervent birder has erected dozens of houses, painted bright red, in the trees overlooking the water. Later, they appeared at our place again—the parents, perhaps, taking the young on their maiden flight.

This year, a chickadee couple seemed to be building a nest in the box on the arbutus tree but something must have frightened them—or else they found a location more to their choosing. There has been a pair around this summer, though, appearing suddenly in clematis or perched briefly on a wire; maybe it's the same couple, raising their brood in a tree cavity somewhere in the vicinity of the house. We hoped they'd choose one of our boxes to nest in but all we can do is make sure each is ready, the cedar sides weathered to silver, each roof intact, and wish for the best.

Platanus orientalis
Raven Libretto

> Xerxes, who chose this way, found here a plane-tree so beautiful, that he presented it with golden ornaments, and put it under the care of one of his Immortals.
>
> —Herodotus, *The Histories*[1]

I was driving down the Coast to do some shopping, and the car radio was on. It was tuned to the CBC's *Richardson's Roundup*. Bill Richardson introduced a piece of music briefly, saying its performer was the American countertenor David Daniels. I'd heard of countertenors, I think, but had confused them with castrati. Certain pop singers used a falsetto voice and I wondered if that was similar. Anyway, nothing — knowledge or ignorance — could have prepared me for what came next. One minute I was driving on an empty highway and the next I was sitting in the car by Homesite Creek, crying into my hands, ravens assembled in the trees above me.

Bill had chosen to play "*Ombra mai fu*" from David Daniels's *Handel Operatic*

Arias, with the Orchestra of the Age of Enlightenment, an ensemble I have since grown to love for the clean pared-back austerity of their sound. But that day I only knew I was hearing a voice and an aria that pierced right to my heart. I knew so little about music, only a kind of blind joy when hearing Bach, sorrow when listening to a Górecki symphony, and wistful nostalgia as the Chieftains played their jigs and laments. But this aria reached down deep, bringing forth those tears, and something else: a desire to know more about singers.

I ordered the CD and listened to it over and over. I loved every piece on it. "*Ombra mai fu*",[2] of course — that love song to a plane tree. In my travels through Europe, I'd grown accustomed to squares in the middle of towns, a single plane tree, or perhaps several, providing leafy shade on hot afternoons. Often a few benches were arranged underneath the trees and invariably someone sat with a bag of shopping at his or her feet, face lifted to the cool leaves. In Greece, old men played backgammon on tables pushed as far into the shade as possible, glasses of ouzo at hand. We stayed several times on the Left Bank in Paris and walked through one such square on our way to shop on the Rue Moufftard. Tables from one little bistro were set with gay cloth and crockery under two big plane trees. I thought of those trees as I listened to the aria.

> *ombra mai fu*
> *di vegetabile*
> *cara ed amabile*
> *soave Più*
> (Never was the shade
> Of a plant
> More dear and lovable,
> Or more soothing.)[3]

And the others: "*Cara sposa, amante cara, / dove sei*"…("My love, my dear betrothed, / where art thou?") The riveting "*Dall'ondoso periglio / salvo mi porta al lido*" ("From the peril of the waves / I have been brought safely to the shore").

I bought some of the operas from which the arias had been extracted: *Serse, Rinaldo, Giulio Cesare in Egitto*. A dear friend with great musical knowledge watched approvingly and sent recordings of other countertenors or entire operas in which I'd expressed interest. Handel and Purcell, music of all sorts sung by Daniel Taylor, Michael Chance, the Deller Consort, and much more: Dowland lute songs, Bach cantatas, Benjamin Britten's folk song arrangements for low voice (I was learning how ravishing the minor chords can be).

    It was thrilling to listen to this music in my quiet house, the volume turned up, the voices filling the space between floor and ceiling, mind and heart. I was nearly fifty years old, and I'd somehow always imagined that opera really meant sopranos. I'd only ever really heard the high voices before — Kiri Te Kanawa, Maria Callas, Renée Fleming. Soprano voices have a way of claiming the space, soaring to those high Cs and beyond, bright and brilliant as filigree. I confess I am exaggerating a little and in any case my ear was so green. So yes, I'd heard tenors, baritones, the low melodic alto voice of Kathleen Ferrier; and when attending operas, I'd enjoyed the duets between the soprano and the lower male voice leads. But here, in my sunlit house, I was listening to something that interested me even more: countertenors and mezzo-sopranos. The range was warm and dense with possibility.

    Once, in Paris, I'd listened to a beautiful countertenor from Martinique sing Pergolesi in Saint-Séverin, the wooden pillars (the one behind the altar shaped like a tree) and ancient pews providing a rich soundboard for those devotions. I was drawn to his voice, recognizing something completely new to me; but I hadn't known how to proceed with that insight. I returned to our hotel on the Left Bank, past the square with its grove of plane trees, the bistro tables chained to a tree, and a man asleep on a bench, newspapers wrapped around his torso. I hummed what I remembered of what I'd heard, and for once I didn't have to stretch for high notes.

    When I was in high school, I sang in our choir. I loved singing. I wasn't particularly good at it but there was the moment, particularly when our choir practised a madrigal, when I could hear the voices braiding elegantly together and knew I was part of this effort; the moment when the puzzling notation made sense in a way that mathematical equations seldom did.

I'd love to have continued with the choir but I didn't get along with its master. He was a diminutive man who wore his hair combed straight up to add an inch or two to his height. He was fussy, his mouth pursed like the anus of a cat, and he preened in front of us during practice. He didn't like me and I didn't like him. I admit now I had attitude — I was sixteen, after all — and was easily distracted. But the occasions when we performed and our harmonies were true and clear were as lovely to me as anything in those years. I never received a mark higher than a C in the choir, but given his knowledge of music and his ability to develop true skills in many of his students, I wish my relationship with the choirmaster had been different.

I began to tentatively hum along to certain arias, after figuring out how the recitative worked: it followed the patterns of speech, the contours of the spoken voice, it seemed to me, and wasn't dependent upon musical structure exactly. The recits advanced the dramatic action of an opera's narrative and the arias opened up the emotional or lyric possibilities of the drama. I entered this musical territory as a complete neophyte, trying to make sense of it on a remote acreage on the Sechelt Peninsula, music playing as I went about my daily tasks. I'd pause in the kneading of bread dough to hear James Bowman in Handel's wonderful oratorio, *The Choice of Hercules,* thinking how smoky and rich his voice in the late recitative, "The sounds breathe fire," and the following aria, "Lead Goddess lead the way." I hung out the laundry, listening to Montserrat Caballé sing "*Chi il bel sogno di Doretta*" from *La Rondine* (*The Swallow*) while all around, the violet-green swallows whirled and dipped.

I loved Handel, finding in his compositions a grand and stately sweep, a generosity to the human voice. The ornamentations weren't just musical acrobatic manoeuvres, but provided natural moments for the singer to engage in something gorgeous and somehow humane — dramatic strength in service to lyrical beauty.

I was recognizing how suited a countertenor voice was to Handel and

Purcell, so too the mezzo-soprano voice. In reading I've done, I've come to understand there are opposing schools of thought (or belief) about role assignments in Handel's oratorios and operas. It reminds me of similar arguments about staging Shakespeare — the gender changes, the debates on the appropriation of voice.

Should a countertenor sing a female role straight or in drag? Should a woman play a man — the trouser roles — as a man, or should her hair tumble down to show us her true nature? Was it Auden who spoke of the tyranny of the pronoun? Never mind. It was fascinating to hear everything: countertenors singing the Sorceress in *Dido and Aeneas*, mezzo-sopranos singing Orfeo in Gluck's divine opera of the same name. The music wanted what it wanted — a voice to enter a role, to caress it, claim it, offer it to a woman in a house in a remote forest, leaning on a broom, in tears.

My hummings became a little less tentative. I'd peer at the small print of the librettos often included with CDs, and attempt to sing. Well, to be honest, I croaked. There was something distinctly raven-like in the sounds that came from my throat. This came as no real surprise. I was surrounded by these birds. On my daily walks, humming and trying to sing, their music was as much a part of my life as that of the opera singers I was listening to.

The ravens are engaging in their vocalizations, sitting in a tall cedar and speculating on the human world below them. *Croanq? Klook?* There is a soft *krrrr* and a long watery gurgle that could go on for some bars, the improvisation of a skilled coloratura, shading and embellishing the notes. And the *tok, tok, tok* — a sound I can mimic by pushing and flattening my tongue up behind my teeth and striking it against my palate. When ravens fly past as I walk over the mail or else dig in the vegetable garden, I make this sound. Almost without fail, the birds do an about-turn, turn on the wing and fly over again, heads quizzically cocked. *Tok, tok, tok*, they'd reply, and wait for a response. We have a brief interchange — I hesitate to call it a conversation, though certainly sounds are made, mimicked, on both sides. Once they determine I'm not a bird, or maybe just because I bore them, they go back to what they were doing. Which often seems to be looking for trouble. Or roadkill.

I've noticed in our area that ravens use the Sunshine Coast Highway as a food lane, and it's not surprising that such intelligent birds have figured out the patterns of squirrel and raccoon deaths. One day there will be a dead animal on the side of the road, and within a very short period of time, a gang of ravens will have cleaned up the mess, muttering and squabbling as they do so. Often a scout will fly low to the pavement, scanning the edges of the road; if I wait long enough, I might see the same bird (or at least I think it's the same bird) returning to the area above the Malaspina substation where there's a huge roost. It will yell as it flies.

In his remarkable book, *Ravens in Winter*, Bernd Heinrich suggests that ravens have a series of sounds used to invite others to share their food, the kind of behaviour that humans would do well to emulate. (An aria from *Theodora* as the table is set, the platters laid out for all to partake.)

Sometimes on the stretch of highway near my home, a deer struck by a car will stagger to the roadside or just into the woods to die. Not long after, I've heard the specific raven yelps that bring other birds. Although I have no hard data, I assume, with the confidence that comes from a long residence in a place, that a message has been sent out to announce that a carcass has been found and that feasting can commence. That's certainly what happens, in much the same sequence, time after time. Now that there are coyotes in our area, and more recently, wolves, I think that interspecies cooperation will develop as the teeth and claws of the mammals prove to be an efficient way to open the bodies; once the wolves and coyotes have eaten their fill, the ravens can come to finish up.

When we walk up behind the Malaspina hydro substation, we often hear the ravens at the roost, yelling and uttering a sound nearer to a yelp. This area is home to a herd of Roosevelt elk. In calving season, I have to wonder if a vulnerable elk calf has been spotted and the alert has gone out to all members of the team. Sometimes we find coyote scats composed of fine beige hair, as though from the young body of a calf. Little slivers of bone, those delicate ankles.

These are stories redolent with operatic pathos — a bull elk gathering his terrified harem as a chorus is sounded closer and closer, a single young female yearling left on her own as the others flee to safety in the dense understory. Think of her in her golden brown summer pelage, eyes filled with the sight of the low bodies of wolves as they approach, aiming for the throat. And the whole while, the ravens making their own dark commentary, the woods loud with their gulps and chortles, a particularly sinister chuckle. "Fly, fly, my brethren, heathen rage pursues us swift, / Arm'd with the terrors of insulting death."[4]

Because "*Ombra mai fu*" consumed my imagination, I wanted to know more about plane trees, remembering their occurrence and shade in the cities and towns I'd visited in Europe. And it seemed that once I was alerted to them, they began to make their presence known in unexpected places. I was reading Virgil, *The Georgics*, and there they were!

> He set in rows his elms when well along,
> Pear trees already hard, and blackthorn sloes,
> Planes large enough to offer drinkers shade.[5]

I remembered the square with its waiting table and how we often found such places ourselves to drink a glass of Prosecco on our way back to our hotel. How the heart longs for green shade in hot climates as the heart longs for music in a quiet house.

A specimen of *Platanus orientalis* grown from seeds gathered at Thermopylae in 1802 flourishes in the Fellows' Garden of Emmanuel College, Cambridge. This is no homage to Xerxes, whose army suffered disproportionate losses at Thermopylae by the Greeks, in part because Xerxes had been so distracted by love for the tree, first encountered by the River Meander, that he didn't keep up the necessary military vigilance. (Or so Herodotus tells us in the *Histories*.)

Another magnificent plane tree of this species grows on the island of Kos and is purported to be the tree under which Hippocrates lectured. Imagine for a moment a generation of young physicians listening to their teacher, their faces dappled with sunlight filtered through the broad green leaves. Imagine one of them idly running his thumb along the prickly surface of a seed vessel, wondering about anatomy and the detachment with which his teacher described his belief that diseases had to do with the environment, with wind and water and weather, and his theories concerning the insufficiency of air experienced by those suffering from epilepsy.[6]

Plane trees have been around for a long time. They exist in the fossil record from the mid-Cretaceous period, roughly 100 million years ago, looking surprisingly like their contemporary selves.[7] They grow well in hot climates as well as harsh climates — the Persians call them *chenar*, the Americans, sycamores. Their leaves are shaped like goose-feet, or (more whimsically) like a map of the Peloponnese.[8] I liked that they were described in terms of both a bird and a geographic location, as though everyone would have a clear image of each in their minds.

On my daughter's recent trip to Greece to study classical archaeological sites, I asked her to bring me home a leaf from a plane tree growing at Thermopylae, wanting that connection to the aria I'd grown to love. She wasn't sure which trees they were and brought instead a tiny fragment of *Lapis lacedaemonius*, the stone from which the monument to the three hundred Spartan warriors was built.

The famous London plane is believed to be a hybrid of the American and Oriental varieties, perhaps the result of a natural marriage between trees from

Greece and America in the Lambeth garden of John Tradescant, gardener to Charles I. I remember the plane trees of Berkeley Square, their plated bark, and the renown they achieved for being able to withstand pollution — those dense coal fogs of London — though when I saw them as a young woman working in Wimbledon and wandering the leafy boroughs on my days off, the air was considerably cleaner. In those years I attended concerts at the Royal Festival Hall, waiting until the last minute to buy tickets in the gods (the cheapest seats or even standing room in the very rafters of the hall) — Janet Baker singing Handel in her rich mezzo voice. Passing the plane trees as a twenty-year-old, I never dreamed I would want to sing of them one day.

Emperor Caligula made a dining room within the confines of a plane tree's trunk, one of those ostentatious things that one somehow expects an emperor to do.[9] Not unlike Xerxes, perhaps, hanging the branches of his tree with golden ornaments and appointing one of his Immortals to care for it for its lifetime. *Cara d'amabile, soave piu*...In his *Sylva*, the esteemed John Evelyn remarked of the plane tree, "Pliny tells us there is no Tree whatsoever which so well defends us from the heat of the Sun in Summer; nor that admits it more kindly in Winter."[10] The more I read, and remembered, the more I realized that the plane tree was rooted in our cultural history as firmly as the olives and oaks I already knew and revered. A heart could be bound to its boughs and leaves as mine had been to the rough bark of the Garry oak in childhood. And the more I read, the more I wanted to sing the aria which had taken me to its shade.

I don't know why it took so long but I finally realized I could take voice lessons and see if I might develop not only my singing voice, but also my knowledge of music. All the years I was raising my children, it wouldn't have occurred to me to indulge myself this way; there wasn't time and not much extra money. But the summer before I turned fifty, I asked a few people for recommendations and was told that I would learn a lot from Shelley Dillon, and that I would love her too. I had met her once, at an event for Earth Day in Roberts Creek: the Goddesses Concert. Women performers had been invited and were taken to a table with a sign: "Reserved for Goddesses and their escorts." (This should happen to every woman at least once in her lifetime, the opportunity to lead her husband or significant other to such a table . . .) I'd been invited to read my poems, the only non-musical Goddess of the evening. Shelley was there with her singing partner, soprano Jo Hammond. I remember their performance vividly—Jo's high sweet voice and Shelley's lower, rich one.

So I called Shelley and set up a trial lesson to see what might be done. Her studio was around the back of an attractive low house and overlooked a lovely garden full of azaleas, magnolias, vines over a twiggy pergola. *Di vegetabile!* Birds enjoyed the feeders and the shelter of the trees. There was a grand piano on a pretty Persian carpet, a stand for music, a waiting area green with plants. A red metronome. Folders containing sheet music were laid out neatly on a shelf. In addition to her voice students, Shelley taught piano. I was nervous.

So much in our culture requires us to maintain the privacy of our mouths. We don't bare our teeth unless we have a good reason or are with intimates; we don't show our open throats; we keep our voices low. I felt shy about following the simplest of instructions: to lower my jaw to make more room; to raise the soft palate; to do an exercise for the correct placement of my breastbone. Bursts of breath through pursed lips. Contorting my facial muscles. And the scales! Oh, I was embarrassed at the sour notes that came from my mouth, that suddenly disappointing orifice. But Shelley was so kind and so helpful that I found myself arranging a regular schedule of lessons and trying to articulate what kind of music it was I wanted to sing.

I thought that this would be like beginner's piano, and remembered an old friend telling me that her husband had bought her a piano for her fiftieth birthday. She arranged for lessons, somehow imagining she would be playing Mozart within weeks. She was humbled to discover that she would be working on "Go Tell Aunt Rhody" for what seemed like months. Intense desire doesn't always translate to even modest ability.

I thought maybe folk songs would be the way to begin and we started with "Come All Ye Fair and Tender Maidens." My attempts were thin and squeaky. And yet it felt wonderful to sing. To try to sing. For several lessons, we practised that song and once I was praised for adding a grace note. I didn't know what that was, but was pleased beyond what was reasonable. When Shelley asked what I wanted to sing next, I wondered if we might try an aria. That aria. "*Ombra mai fu.*" David Daniels made it sound so effortless. She smiled. "If you were a fifteen-year-old girl wanting a career as a singer, I'd tell you that you weren't ready yet. But I think you'll learn a lot from trying, so why not?"

Shelley played the opening bars of "*Ombra mai fu.*" My heart began to race and I could feel my shoulders tensing. I wanted so badly to do some sort of justice not just to the song but to the woman who sat at Homesite Creek and wept for the beauty of those words in her small car...David Daniels's offering of them, the way they eased into my heart like a homecoming, a blessing for trees and the solace of their shade.

Of course I mangled it. Everything conspired against me, myself most of all. My lack of musical ear, breath control, agility, and support; but I knew I wanted to keep trying. Driving home, I'd sing the scales over and over, attempt the arpeggios — and could manage perhaps three without the comfort of the piano to guide me up and down.

And there would be the audience of ravens as I drove the long highway home, standing on the roadsides with their complicit gazes, the little falsetto yelps as I passed. Sometimes I'd stop the car to see what it was they were doing. There might be a flattened squirrel or snake, depending on the season, but often there would be nothing that I could determine might attract them to stand around as though waiting. As though waiting for me to let them know how the lessons were going.

An Unkindness of Ravens

Ravens mate for life although infidelity is not unknown. They have the largest brain of any songbird. They have learned to make use of sticks and other things as rudimentary tools. Their vocabulary is considerable. The ones we know on the west coast of British Columbia are thought to have come over the Bering land bridge from Asia. Their name, *Corvus corax*, has classical roots: the genus, *Corvus*, comes from Latin; the specific name, *corax*, from the Greek. In Old English, they were *hraefn*, in Old Norse, *hrafn*. These ancient words contain something of the raspy noise of their language! What comes first, a name or a sound? Is this the chicken and the egg riddle? I imagine the ravens waiting for a hen to leave its eggs unattended, and then swooping in to feast on the rich yolk. If a tiny embryo had already formed, so much the better. The ravens certainly wouldn't have been participants in a debate about their origins but might have been seen in the distance, muttering, taking any opportunity for a meal.

The girl from piano

I am remembering that first lesson. While I tried to sing "Come All Ye Fair and Tender Maidens," a little girl slipped in the door of the studio, brushing by the rosemary shrub at the threshold so that the entire room filled with its resiny odour. She sat on the chair in the waiting area, clutching her sheaf of music, watching me with the intense fierce look of a girl who doesn't miss a trick. I know, because I was once that girl too, eyes open to the world and all that it might offer or deny. This little girl had red curls, only one front tooth, and freckles over her nose.

I finished my lesson and spent a few minutes arranging a time for a sequence of instruction that fall. The girl watched, and listened. When I walked out, I said hello to her and her face broke open into a smile that lit the room.

I continued down the Coast to the town of Sechelt where I had errands to do. I felt ten feet tall, full of music, its rich possibilities, the excitement of actually learning something new. I spent an hour buying groceries and then stopped at the library to take out the next week's worth of books. There was the little girl again, this time with a woman who could only be her mother — the same hair, the same bright eyes.

"Hello again," I said, as I took my books to the desk to check them out.

She didn't say a word to me, but loudly whispered to her mother, "It's the girl from piano!"

And I, almost fifty, smiled at them both as I left the library.

When I was the age of Jenna (for I discovered that this was her name), I lived in Nova Scotia and had a friend who was the son of a United Church minister. His family lived behind ours, with a shared fence, easy to climb over. This boy (whose name I forget) took piano lessons and sometimes I'd sit quietly in his living room while his teacher (it might even have been his mother) guided him through scales, little exercises, and even some simple tunes.

I wanted so badly to learn to play the piano but my family moved every two years and our furniture would be driven across the country in a moving van, often with damage incurred along the way. A piano was considered impossible. Neither of my parents knew anything about music and the idea that a child might want to learn to play was frivolous in the extreme. I'd listen to my friend patiently practising, and sometimes he'd show me how to play "Chopsticks" or a one-handed version of "Michael Row Your Boat Ashore." I was taken by the magic of creating music by pressing the ivory keys in sequence: plink…plink…plink…plink…plink. Ha…le…lu…u…jah. Astonishing. And when the boy began to play the first classical piece I ever heard, "*Für Elise*," I cried. I wanted to have this kind of beauty in my life in a way that was so personal and immediate, music unfolding from my fingers like soft cloth.

I was also learning to knit that winter, an untidy mess of knots and lumps that I hoped might become a scarf one day if I was patient, and was struck by the similarity. Starting with a kind of bold confidence, immediately making a mistake. Going back to the beginning, as the boy so often did, and as I did with my yarn, determined that the next time would be better. Taking deep breaths to stay patient as I cast on, knit, then purled, keeping the steps in my mind as clearly as I could. As I hoped, in later years, would happen with singing. I didn't know then that a throat could be trained the way fingers could be, that agility could be practised and learned. That a girl could age in the blink of an eye but still find a way to brush past the rosemary into the studio to open her mouth and hope that what came out might resemble music, though whether the beauty of plane trees or the guttural commentaries of ravens was not yet clear.

> To the hills and the vales, to the rocks and the mountains,
> To the musical groves and the cool shady fountains...
> —*Dido and Aeneas*[11]

I was waiting in the car while John filled the tank with gas from the Extra Foods gas pumps in Merritt. I had been thinking of a few favourite arias, humming to myself as I leaned my face against the window. The day was mild, not quite warm, and we'd driven down from Kamloops where we'd spent the weekend with our older son Forrest; he was living there for the months of May and June to teach Canadian history as a sessional instructor. I was humming and remembering; one so often leads to the other, in no particular order. And then I saw the most extraordinary sight: at least twenty ravens on a dumpster behind McDonald's. Some were perched on the side of the dumpster itself and some were lined up on the surrounding fence. One was on the ground, walking away in a slightly pensive posture. I realized the aria I was humming was Dido's lament from Purcell's sublime opera, *Dido and Aeneas*.

I've only seen one performance of this opera, though I have at least three recordings—one with the late great Tatiana Troyanos as the doomed Dido. The production I saw at the Chan Centre in Vancouver was modestly staged; this is not unusual for Baroque opera, I understand. The chorus stood on risers, the principals moved to centre stage to perform. The singers all wore evening dress. It was an effective performance, the singing was generally good and occasionally quite fine, but I missed the splendour of full operatic spectacle.

The ravens on the fence were different sizes. Some of them would lift their wings briefly and then settle again. One large bird sat on a pole — a sentinel? I got out of the car and walked towards them. The sound was amazing. Chortles, gurgling, clicks, toks. The raven on the ground stood still and posed in profile, throat shaggy as a ruff, while others plundered the dumpster. I listened. I know that the avian vocal organ is the *syrinx*, a bony structure at the bottom of the trachea. In humans and most other mammals, our vocal chords are contained within the larynx, located at the top of the trachea. As air is blown across them, the chords or folds vibrate to produce sound. Well, as I understand it, vocal chords oscillate as they make contact with each other.

Birds have a particular skill, or at least some species do — song thrushes for instance, though I'm not certain about ravens: they can control two sides of their trachea independently, thus producing two notes at once. A throat that is its own duet! When I have difficulty even creating one true note, I think of this with something like envy, something like awe.

"Split-voice, wind-carried child of sound..." sang Theocritus of the *syrinx* in the third century BC.[12]

It was a kind of magic, a black magic, that I was in a part of the country I love — the golden hills rippling like buckskins, rainbows appearing over the irrigation sprinklers on the hay fields, cattle with their young waiting to be transferred to the high ranges — watching a theatre piece at the McDonald's while all around me drivers filled their vehicles with gas, shoppers transferred groceries from carts to the backs of pickup trucks, someone swept the sidewalk in front of the Dollar Store, the *Open* light at the Taco Del Mar was flashing red and blue, and a loud clanging of iron up at the Canadian Tire indicated someone was hard at work with wrenches and jacks.

The ravens on the fence muttered and yelped. Periodically they'd change places. One would hop down into the dumpster while another would take its place on the risers. It seemed choreographed, orderly. Occasionally there would be a loud falsetto passage, like a countertenor from *Dido and Aeneas* calling for some kind of macabre action:

> Wayward sisters, you that fright
> The lonely traveller by night,
> Who, like dismal ravens crying...

And they did, on their risers, a concerted high cry, pretending to be dismal, though obviously exuberant at the possibility of yesterday's hamburgers and stale sesame-seed buns.

> Beat the windows of the dying
> Appear! Appear at my call, and share in the fame
> Of a mischief shall make all Carthage flame.
> Appear!

The black chorus on the fence would answer:

> In our deep vaulted cell charm we'll prepare,
> Too dreadful a practice for this open air.

I walked closer. A few birds eyed me suspiciously. Was I there to steal from the dumpster? I looked away so they wouldn't get ideas. That one on the pole—a sentinel? Or a conductor? It watched but never made a sound.

> Thanks to these lonesome vales
> These desert hills and dales...

The raven on the ground, followed now by another, uttered its own noise. It gurgled, chuckled as though in great amusement as it gazed out towards the hills rising beyond the highway leading to Quilchena, and then it began to utter a ravishing song, like water falling into more water, rattling against stones. The other bird inclined its head, listening and considering. Although I know it's an anthropomorphic imposition, I am tempted to say that this was Dido's grand aria. The raven was as regal as any queen as it croaked and gulped.

> When I am laid, am lai...aaa...id in earth,
> May my wrongs create
> No trouble, no trouble...
> Remember me, remember me, but ahh...ahhh, forget my fate!

An easy reach to that high G. *Remember me!* And the second bird trailed behind like Dido's faithful maid, Belinda.

I listened, closing my eyes to shut out the Extra Foods store, the Canadian Tire, the Tim Hortons with a line of cars waiting at the drive-through lane, the McDonald's itself. This was music worthy of my concentration, an outdoor opera with a cast in full voice. Even the sorceress's role cunningly sung by a countertenor, which would have pleased Henry Purcell. The falsetto was quite clean and confident, which made me realize how art often sounds artless but the reality is that there are obstacles to such beauty. The Adam's apple lodged in the male throat (and to a lesser degree, the female throat, too, but not as obvious a protrusion) is a thickness of cartilage around the precious voice box. And if I had trouble hitting a G, a clear A flat, how on earth did a countertenor do it? A raven?

When I opened my eyes again, there was just a raven walking across the road to the grassy verge, where a sprinkler spun under a couple of small pines. Spreading its wings a little, tilting its head up and opening its beak, throat shaggy and silent, it entered the spray.

In my mind, I see three plane trees, leafy and beautiful, two men playing backgammon in their shade. From their branches, golden ornaments hang and turn in the breeze. But wait, there are ravens on the boughs too, one there, and another there, and look, two more on the crown. And wait, listen: the ravens are singing *"Ombra mai fu."*

I read about her before I heard her sing. There was a wonderful profile on Lorraine Hunt Lieberson in *The New Yorker*, written by Charles Michener. After reading it, I thought, This is a singer I must listen to. I wasn't sure why, but I think it was because she didn't sound like a diva in the profile. She'd done things other than sing; for instance, she had helped to erect living quarters in the yard of a Mexican prison so she could be near her boyfriend of the time, a man convicted on drug charges. Her allusions to astrology amused me; I remembered my own encounter with a medium who told me I was in the care of Pan.

Lorraine Hunt Lieberson photographed by Richard Avedon—a gorgeous woman with long hair and a shirt open at the neck, smiling. It was so refreshing to see a generous mouth and crinkly eyes instead of a stern-faced Valkyrie in breastplate and horned helmet, or an elaborate brocade gown showing fierce cleavage bound in whalebone. Michener had heard her sing first at a benefit concert at the home of Leonard Bernstein and had been so taken by her voice that he said to her, "'You have one of the most beautiful voices I've ever heard. Who are you?' 'I'm a violist,' she replied, with the trace of a smile."[13]

I love this anecdote for what it tells us about the woman—her voice, the sense of that voice as an instrument, a mid-range instrument, with a depth and texture that one doesn't encounter very often. Not a violin, not a cello. Elsewhere in the profile, Michener refers to the "darkly gleaming" quality of her voice. (I thought of coffee, the kind I love most: dark roasted beans made into a strong infusion, flavour consistent with strength.)

What I heard in Lorraine Hunt Lieberson's voice the first time I listened to it—after reading her profile, I went out to buy whatever I could find:

a recording of Handel arias (of course) and two Bach Cantatas, BWV 199 and BWV 82—was a luxurious depth of emotional engagement. This was not simply a technical performance, a gifted virtuosity, but something lived and felt, singing that was a process, an act of offering.

Once she was on my radar, I encountered her everywhere. In truth, she was there before I knew her name. A recording of *Dido and Aeneas*, bought and listened to reverently: when I took it from the shelf and checked, yes, it was Lorraine singing Dido. Mentioning her to a friend who has sent me gifts of music, to my surprise he told me I was revealing excellent taste. But I have no particular taste, I thought. I have no basis for liking what I like. I only know what moves me. And she did.

I thought of that quality in some Renaissance painting: *chiaroscuro*—a balance between light and dark. The brilliant highlights on skin or interiors serve to effectively contrast the areas of strong shadow. Caravaggio for example: his *Madonna of the Pilgrims* shows us the dark-haired Mother with her child, her throat a chalice of light and her infant's thigh gleaming while behind her a dim plaster wall shows brick, one pilgrim's cloak ripples on his shoulders like a pelt, and the Madonna's own shirt and skirt are studies in dusky tonality. Lorraine's voice has this balance—the clarity and brightness of tone held in exquisite tension with such deep and velvety warmth. You could *see* her singing, although of course it was a recording.

I began my own singing lessons with her as my inner muse. I'd leave my lesson and listen to her sing "*Ich habe genug*" on my way home along the Coast Highway, taking in every note, every passage of open-throated beauty as I drove past views of islands, the Strait of Georgia glittering in sun or gun-dull in rain. I'd never heard this cantata sung by a mezzo-soprano before (though in truth I think the transposition is nearer the alto range) but had heard a bass, a soprano, the gorgeous obbligato passages played by flute. In this recording, the oboe d'amore echoed Lorraine's rich vibrato. Or she echoed the oboe d'amore, the two of them engaged in a long legato duet.

Where did Bach's glorious creation end and Lorraine begin? I'd never known that a singer could reside so completely within the emotional and spiritual landscape of a composer that he or she was somehow its embodiment. (I had heard this in a way with Rostropovich's transcendent interpretation of Bach's suites for unaccompanied cello. But with Cantata BWV 82, in this recording, it was as though Lorraine and the oboe d'amore and the sorrowful resignation expressed by Simeon, as he both anticipates and regrets his own impending death, are all of a voice, all in the same place.)

Eventually, I began to sing along with Lorraine. Having no piano at home, no means to practise in the usual way, I taped my lessons and practised by singing to my own exercises. I'd cringe as I pressed Play and heard my thin efforts, marred by missed notes, no sense of timing, my voice straining to sing even a D. But I'd persevere, trying to improve.

Discovering that I could practise by singing with Lorraine was an unexpected blessing. Not the cantatas, of course, or the Brahms, the Schubert. But there were several Handel pieces — "*Ombra mai fu*" for instance, the plane tree shining and beloved — that I had also begun to learn and there was no better model than her recordings. I'd like to believe I learned phrasing and pitch from Lorraine, and how to enter a song with my whole self, giving myself over to it. I loved the sound of her voice in my empty house, the fully open "As With Rosy Steps the Morn" ringing up the stairs to my bedroom where sunlight filtered through the white linen curtains, her "Deep River" like a block of dark chocolate.

A Beginner's Repertoire (annotated)
1) "Come All Ye Fair and Tender Maidens": I found a Joan Baez songbook and Xeroxed two copies of this, one for Shelley and one for me. We worked on it for about three lessons and I found it very difficult. But there were moments when I'd feel my throat open and a phrase would sound out with something like a ring. I knew I wanted to do this for the rest of my life.

2) I've already confessed to mangling "*Ombra mai fu*," the aria that led me to singing lessons in the first place. I'd been listening to Daniel Taylor's *Portrait* CD and was very taken by the John Dowland song, "Flow My Tears." I loved its mournful quality, its elegance. "Can we try this?" I asked. And Shelley kindly said yes, though I realize now what a trial it must have been for her to guide my weak and faltering voice through such a gorgeous piece. I'd take huge lungfuls of air and attack the song like a drowning woman, seeing the E looming towards the end — "Hap-py, hap-py they that in hell / Feel not the world's de-spite" — and feeling my heart flutter with anxiety. Could I reach it? Would I strain and not arrive? Whew. But I practised regularly, my tape deck set up in the corner of the kitchen, trying it over and over again. And now, years later, I wonder how I ever could have chosen that song with only a month or two of lessons behind me. It is so hard and requires such breath control.

3) My older son had been to Lester B. Pearson College of the Pacific and had sung in the college choir. One piece he'd enjoyed was "Jerusalem," stanzas from Blake's "Prophetic Books" set to music by C. Hubert H. Parry. After hearing him sing it in his light tenor, I tried to entice him to teach me but he wasn't willing. Sometimes I'd hum when I heard him singing but wasn't sure of the melody, the lyrics. So Shelley ordered the sheet music and we laughed our way through its very hymn-like cadences. (As a child, I sang in a church choir and enjoyed the practices, the occasional appearance before the congregation in our dark blue gowns with the contrasting white yokes. But it was hard not to laugh when the [very] senior choir sang "Bringing in the Sheaves" or "Holy Holy Holy." I'm sure they felt wonderful — well, I know they felt wonderful and all the recent scientific evidence suggests that they were unknowingly keeping their blood pressure down, their serotonin levels up, their immune systems in healthy readiness — but they sounded so frail and so trembly.) Still, I'd get goosebumps when we came to the robust final stanza: "Bring me my bow of burning gold! Bring me my arrows of desire!" (How did arrows of desire fit in a poem about the Holy Lamb of God? Well, that was Blake, I suppose — heaven and hell, innocence and experience.)

4) "Why not try this," she asked? I'd bought *Italian Songs and Arias*, arranged for medium voice, and we were leafing through it. "This" was "*Caro mio ben,*" ah dearest love, by one Tommaso Giordani. What I liked about the collection as a whole was the way each song was presented: a page of background information, source, bibliography should one want to follow up, and a phonetic guide to pronunciation. There was also a CD of accompaniment included for those like me who had no way to practise the piece at home. And what I liked about "*Caro mio ben*" itself was its accessibility, both as a song and as an exercise for a beginner like me. It wasn't too high (I was learning about tessitura and although there were some high notes in this piece, the portion of the range that was characteristic was very manageable). I had several months of Italian at university, just enough to give me a little bit of confidence if the tempo was, as with this song, larghetto. There were passages with interesting ornamentation, too, to aspire to. (I'd listen to Cecilia Bartoli sing this song and was breathless at her quicksilver grace as I followed the score.) This was a song I revisited, every six months or so, to see if I'd improved enough to do it any kind of amateur justice.

5) "Could we try 'When I am laid in earth'?" "We can, of course." I muddled through the recitative, though finding it easier than "*Frondi tenere,*" which I never sang after the first few attempts. And then the exquisite cry from the heart of Dido. Or it would have been beautiful if my throat hadn't closed completely for the G above the staff. I was so disappointed with myself. Shelley very sweetly took the final refrain down half an octave but I was completely aware that I was mangling a thing of great beauty. "It's not that it's too high for you," she explained, "but that you have to make a leap to a vowel, to an eeee, and you don't want it to sound like you've seen a mouse. It's always difficult."

6) I discovered I loved Benjamin Britten's folk song arrangements and that I could sing them reasonably well. "Down by the Salley Gardens"—a poem by

William Butler Yeats, haunted by a long-ago love; "Last Rose of Summer" with its dark embellishments. Shelley had this transposed for me as the arrangement was for soprano voice; after three years of lessons, I could almost manage, apart from several ornaments which went to high C. Almost, but not quite. And the misses were dreadful. And I also was intrigued by the American song-collector and composer John Jacob Niles. I worked on "Black is the Colour," a song I couldn't forget and found myself singing mournfully while working in the garden or washing dishes. There was something in it of Dowland, something of Britten—the minor chords, the flats. The wonderful Daniel Taylor sings this with lute accompaniment by Sylvain Bergeron on the splendid *Lie Down, Poor Heart,* and I learned a lot about timing from him.

7) "Amarilli," "And So It Goes," "Aye Fond Kiss," "Fairest Isle," "The Cuckoo," "Since First I Saw Your Face" (which took me full circle back to that choir when this had been a piece we sang in four parts. It made me wistful for missed opportunities, bad timing—maybe I could have become a singer in those sad old years in high school), "Sometimes I Feel Like A Motherless Child," "*Laschia ch'io pianga,*" and "Where'er you walk" (I love singing Handel!), "*O Leggiadri occhi belli*" (my Italian stumbling to keep up), "Have You Seen but a White Lily Where it Grows?", more Niles ("The Black Dress" and "The Gambler's Lament," which I sang with rib cage expanded and my tentative vibrato, but I longed to put aside my small classical accomplishments for these to sing them as plaintively, wistfully as Emmylou Harris might).

8) After the second year, after many exercises, arpeggios inching up a half-tone every few months, we tried "When I am Laid in Earth" again. And I could do it! I bought myself a deep red taffeta evening skirt at my favourite second-hand clothing store as a fitting reward. When I wear it, I am Dido, collapsing into her handmaiden's arms.

When I learned of Lorraine Hunt Lieberson's death, in July 2006, at the young age of fifty-two, I walked around numb and distracted. How could it be: a young and accomplished woman gone to spirit so soon? This was someone I'd wanted to have in my life as surely as my family, a guide through the difficulty of scales and arpeggios, someone to sing Bach with, the generous songs of Handel. And yet how could I mourn someone I'd never met, who had a husband and family of her own who would be inconsolable, burdened with sorrow. My grief seemed a remote intrusion, an affectation. I did what I could. I located every CD I could find so that I'd have the solace of her voice. The poems of Neruda and Rilke her husband, the composer Peter Lieberson, had set for her. The Spanish love songs she sang with Joseph Kaiser.

I read everything I could about her, wanting the company of her name spoken by others. I knew that she had been scheduled to sing Orfeo in the great Gluck opera at the Met (that story of the mourning lover descending with great courage to the underworld to beg the return of his beloved) — one of those marvellous gender-bending tropes, the ravishing mezzo-soprano singing the male lead (I think of Tatiana Troyanos, Denyce Graves) — and was surprised and thrilled to see that David Daniels would sing the role instead. In an ideal world, I'd have begged, borrowed, or stolen to be there, watching the performance (a Mark Morris extravaganza), but know somehow that Daniels's Orfeo would have been sublime and I'm content to have it so in my imagination. Orfeo's arias — "*Che puro ciel*" as he wanders Elysium:

> What pure skies, what bright sun,
> what new clear light shines here?
> What sweet enchanting harmonies
> are created by this blend
> of the song of the birds,
> the murmur of the streams,
> the whisper of the breezes!;

and the gorgeous *"Che farò senza Euridice?"*:

> What can I do without Euridice?
> Where can I go without my beloved? — [14]

are tender and tragic.

In Gluck's *Orfeo ed Euridice*, there is a happy ending. Amor allows for the couple to be reunited, though the great mythographers assure us that Euridice was lost to Orfeo forever, music without its muse. How fitting that the production was dedicated to Lorraine. For me, for so many, having found the voice of our dreams, we must adjust to absence now she is gone to the Elysian fields (I imagine them to be like the place described in Theocritus's *Idylls*, a place where

> Tall pines grew close by, poplars, and plane-trees,
> leafy cypresses; and fragrant flowers, thick in the meadows,
> a labour of love, as spring dies, for rough bees…[15]

that haunted place of perfect peace).

Predators, tricksters, comics, monogamists, careful parents, scavengers, demiurges, shape-shifters, opportunists, acrobats on the high currents of air, practitioners of song. If I am honest, I must confess that ravens are almost as influential to my sense of music as the other singers I have learned to love. How they croak and gulp in the tall firs on the Malaspina trail, in an area I think must be their roost. How they can klook and trill, utter two notes at once from throats shaggy with feathers, a dark and original duet with the self.

I have longed for the clear voice of a mezzo-soprano, to sing perfectly of plane trees and their gift of shade, but am too careless and erratic to train to that level. I say I began too late, but I suspect I wouldn't have had the tenacity or the self-discipline even in my youth to repeat those scales, those exercises to improve tone and dexterity. An afternoon on a single vowel, a year, a decade, learning to match vowels to pitch? Coaxing and adjusting my vocal tract for perfect reactance? Oh, I wish it was easier, and that I didn't love too many things at once. Distracted from practice to water plants, to pollinate the lemon tree in the depths of winter when no bees were in the sunroom to do it, to look up a reference, to design a quilt, to read a poem by Sappho, to call a far-flung son to ask about Christmas or the physics of acoustics. To sit by a window dreaming of Paris and its churches, its squares set with tables under plane trees, while here the cedar boughs droop with their weight of birds.

And it turns out I want something else of singing too, not formal or professional: I want, yes, I want the spontaneity of stepping out from a parking lot in the golden grasslands south of Kamloops to watch a cast of ravens in the open air singing their own opera of the everyday while cars pass on the roadways and shoppers load groceries into their trucks for the long drive home.

Pinus ponderosa
A Serious Waltz

> The heart is the warmest organ. It has a definite beat and movement of its own as if it were a second living creature inside the body.
> — Pliny, *Natural History*

There is a moment, driving on the Coquihalla Highway between Hope and Merritt, when the landscape changes. The spruces, which have carried snow in their strong, supple arms, give way to ponderosa pines. The highway begins its long descent to the Nicola Valley where the pines stand among the soft grasses and artemisia, providing shade for cattle, seeds for chipmunks, Clark's nutcrackers, grey jays, and the myriad small birds that dart in and out of the branches.

Is it excessive to say that I have loved those ponderosas all my life? On family trips in my childhood, up the Fraser Canyon — this was before the construction of the Coquihalla Highway — I remember waiting for them. Was it around Boston Bar where the beautiful groves began? Or at least individual pines on the benches above the river, standing with the firs and delicate aspens. The pattern of branches against the sky changed as the firs and cedars gave way to the lyrical pines, their airy latticework and straight trunks a signal that we had entered new country. Sometimes we'd stop north of Lytton at the picnic site at Skihist and there were pines in that dry air above the river, stately and sweet smelling. I loved their bark, thick and puzzle-shaped, and would scrape little drops of sap off, rolling it between my fingers. In my sleeping bag at night, in a tent of blue canvas, I'd sniff my hands for the faint memory of resin, of vanilla.

In that tent, patched by my father with scraps of brown canvas cut from even older tents, we slept our way across Canada, by rivers snaking through prairie grasslands low with wolf willow or slender birch, in boreal forests, the mixed forests of the Great Lakes, St. Lawrence, and Atlantic regions; but it's British Columbia's park-like ponderosa forests that I remember with an affection approximating deep attachment.

In the other woodlands, mosquitoes rose in clouds; I was allergic to them. Once we had to go to the emergency room of an Alberta hospital after camping in drizzle somewhere low and marshy; my infected mosquito bites were dressed with gauze, and I was given penicillin. (I can still feel the fierce sting as the bandages were changed each day, the tape drawn quickly off, taking the fine hairs of my arms and legs with it.) But in woodlands graced by pines, it was usually dry. Instead of rain, we'd wake to the sound of squirrels and grasshoppers clicking in the grass. There was almost nothing nicer than the smell of coffee perking in the aluminum pot; it came to me filtered through fire, bright in a circle of stones, as I pulled on my shorts, dancing a little because I had to pee. Our tent was gilded in sunlight and pollen, surrounded by brown-eyed Susans and rabbitbrush. We'd cut a little clump of their flowers for the picnic table, keeping it fresh in a Campbell's Soup can. This was beauty unrivalled, not even by a hothouse bouquet in a crystal vase or an arrangement of orchids in a Japanese ceramic dish.

Those waking moments, when the landscape changes, are portals, green with branches. Passing into ponderosa country, I remember to take a small twig or a cone. I keep the cones in a bowl at home, a way back as sure as dreaming. The cones are arranged and rearranged. Stray seeds adrift on the bottom of the bowl, as though their secrets might be revealed in the right presentation: a divination.

Late September 2008 — I'm determined to learn to make pine-needle baskets. I bought one for John for Christmas a few years ago and have come to love its sturdy beauty — brown needles still smelling faintly of their origins woven into a bowl, stitched with some fine thread. A top fits snugly, a wooden button plain in its centre. I've read whatever I can find on the baskets, have downloaded instructions from a Web site devoted to Northwest baskets, and ordered one of the books recommended for further study.

The materials needed are pine needles, raffia, and number 18 tapestry needles. The latter two are easy. The dollar shop in Sechelt has the raffia, even the recommended type: not fireproofed, which apparently makes the raffia waxy and slippery, and difficult to work with. I find the tapestry needles at the sewing shop in the mall, a whole range of sizes, all wide-eyed, with blunt tips.

So it's the pine needles themselves for which I will have to go farther afield. I know there are a few mature white pines (*Pinus monticola*) in Halfmoon Bay — I've brought home their elegant cones to decorate Christmas parcels of homemade jams and fruitcake. But at ten centimetres, their needles are too short to turn and wind into coils. If I lived in Florida, I could choose from longleaf (up to forty-six centimetres long, these produce the longest needles of the native North American species), loblolly, slash, and pond pines. In other western areas (but not close to home), there are Jeffrey, digger, Colter, Apache, and Toray pines. On the west coast, though not as far north as where I live on the Sechelt Peninsula, and in the interior of British Columbia, it's the ponderosa that is most widely used for basket making, with needles ranging from eighteen to twenty-three centimetres long. That settles it. A run up into the Nicola and Thompson valleys is necessary.

Monck Park is pretty deserted when we drive in early on a Thursday afternoon. Two campers set up down by Nicola Lake, signs of abandoned activity in the parking lot—a backhoe, a wheelbarrow, a broken pad of concrete with the informational sign it supported lying on its side in the grass. Under the large ponderosas fringing the parking lot are heaps of fallen needles. Masses of them. I've brought a large yellow detergent bucket from home and hope that by filling it with needles, I will have enough for my baskets. I begin to gather them by the handfuls. The smell is wonderful, though truly it isn't just dry pine needles but also sage (*Artemisia tridentata*) and its smaller, woollier cousin *Artemisia frigida*, or pasture wormwood. Brushing my hands over their flowering tops releases a pungent, lemony odour that tells me where I am more accurately than any map.

For many summers, I camped here with my family and watched my children play among these plants. They ran among the pines and came to the campsite with sticky hands from the sap, their skin golden with pollen. We made fires with dry needles and logs of fallen ponderosa, which snapped and burned hot with pitch. I folded my children's clothing at bedtime, their socks pierced with seeds. When I tucked each child into a sleeping bag and bent to kiss them goodnight, I smelled lake water and smoke in their hair.

Now my children have moved far away, where other trees watch them sleep—an apple tree and yews overlooking Cordova Bay outside Victoria; beeches and butternuts on a leafy street in London, Ontario; "something with green leaves, and another with purple ones" in Toronto (reported by a son unconvinced of the importance of specificity, at least in botanical matters).

A day or two later, I am under some dead ponderosas near Heffley Creek. We've stopped in a recreation area on the road up to Sun Peaks, mostly so we can drink our coffee and eat something because we drove away early from Quilchena without breakfast, planning to stop at a diner for a proper meal of eggs and sausage, but didn't find a likely place on the drive north on the 5A.

All around us, as far as the eye can see, there are standing dead trees—and very few living. I muse about the living trees and the dead, for these trees are not dead as a result of age or a falling meteorite or gases from industry. Rather, the mountain pine beetle (*Dendroctonus ponderosae*) has seriously decimated the ponderosa pines of the British Columbia interior. This has been happening for some years now and has filled the business pages of the newspapers as the lumber industry faces the challenges of declining healthy timber stocks and rising fire hazard as the woods become stands of pitch-infused tinder.

Science writers following the accelerated pace of global warming are breathless with numbers and charts. When I flew over the area last, returning home from Ontario, I was shocked at the patches of red in the dark green forests below. By the time I was over the Interior, the plane beginning its descent into Vancouver, I watched for the river systems, forests, cities, and small towns, all very distinguishable from the little windows. I've often followed our route home in my heart by tracing the Fraser River below with my finger on the glass. Anyway, I was startled by the way red was overtaking green in so many areas. Of course, I'd noticed this driving through the Interior, but some areas were still untouched, it seemed, by the beetles and their devastation. I suppose this was what I'd wanted to remember: the groves of healthy trees standing in their drifts of golden grass. I'd seen them in morning light and evening, moving with slow dignity in the wind, as though waltzing—a modest sway of upper branches brushing softly against the nearest tree.

Beneath the dead trees, thinking about this, I bend to begin gathering needles. I suddenly decide to keep the needles from the various areas separate. I should have thought of this earlier, because essentially I've filled the yellow bucket with Nicola Lake needles, not in any tidy way, and I'm realizing that bundles properly tied with string or elastic bands, all the sheaths at one end, will take up far less room than a haphazard pile of needles put in any old way. I put elastic bands around each end of the Heffley Creek bundle and then label it with a bit of paper from my notebook. It's then that I notice all the little seedlings coming up near where we parked our car.

In one small scrap of ground, perhaps one square foot, there are six tiny pines. And when I look elsewhere, I see that it's the same. A forest in miniature, each tree no more than eight centimetres tall. I realize I must've stepped on many as I drank my coffee, ate my handful of almonds and my apple, and bent to gather the dry needles. I was so single-minded about choosing the longest lengths of needles, and preoccupied by the dead, that I hadn't noticed the tender green seedlings.

The recreation area is open and there aren't any signs to indicate it's been replanted. A dog show of some sort is going on over in the camping area adjacent to where we've parked. We keep seeing people walking their dogs on the ridge just beyond us, high-stepping standard poodles, clipped to elegant topiary, and belligerent Jack Russells. When I look at the area where I gathered needles most actively, I see that the little seedlings have sprung back to their full height. Whatever damage I've done is minimal—a small consolation here, where the dead trees are like mortuary poles. I decide to take a specimen home with me, one of those growing thickly enough that most will die when the dominant seedling overshadows the others. I use my coffee cup and a knife to trace the path of the root (surprisingly long) through the dry earth. I take duff from around the little grove of pines and then wrap the cardboard cup in a plastic bag.

After a few days in Kamloops, we drive west along the Thompson Plateau, stopping at a rest area overlooking Kamloops Lake just east of Savona. When we'd stopped here a few months ago, in May, I remember how the dying pines cut a red swath down the cliffs to the shores of the lake. This time I immediately see that several of the big ones have been cut down and into short lengths, and the bark pried off to reveal the beetle galleries underneath. It's a script, a scrawl of purpose and intention, where the beetles have tunnelled into the phloem layer. Yet a pygmy nuthatch moves along the exposed trunk, from side to side, alert for beetles or larvae. The heaps of needles are long and clean. I collect a bundle, labelling it with the place and date.

> I alight as a beetle on the empty throne which is on your bark, O Re!
> — from an Egyptian Pyramid Text from the fifth and sixth dynasties, ca. 2556 - 2150 BC[1]

Consider the beetles themselves. Black or dark brown, the size of a grain of rice. Not unattractive, as beetles go. Not immediately scary. They bore into the bark of the host tree — ponderosa, lodgepole, Scots, limber, bristlecone, pinyon (these last two less frequently) — and create vertical galleries in the phloem layer, where they deposit their eggs. The larvae feed horizontally; this scribble of galleries and feeding paths is what you can see on a dead tree once the bark is removed, a calligraphy of darker brown against the tan wood. Read these glyphs like a text: a story of soft tissue consumed by larvae as they eat their way into adulthood and out of the tree; the invasion of a type of *ophiostoma* or blue-staining fungi introduced by the adults — it adheres to their exoskeleton and travels with them — which spreads and disrupts the flow of water and nutrients within the tree, causing its death. Run your hands along a standing tree to feel the pitch tubes, the lumps of resin that indicate where a beetle has burrowed in, or out, of a tree, the tree's attempts to heal itself. Smell your hands. Turpentine will come to mind, and the antiseptic whiff of a travelling vet.

It's perverse to admire an insect capable of such devastation. And yet beetles have been worshipped since prehistory. Scarab pendants have been dated to the late Palaeolithic — 10,000 to 20,000 years ago. Because of their ability to fly and to burrow, the beetles were thought of as mediators between the terrestrial and celestial worlds. They have associations with creation in societies ranging from shamanic groups in South America to Sumatra to Egypt. In those ancient cultures which believed that humans were modelled from clay, what better symbol than that ur-potter, the scarab beetle, gathering and shaping dung into a ball? If one observed that ball long enough, it would come to life, the tiny larvae of a new generation emerging from its centre.

Coming down the long slope from Rose Hill to Kamloops a couple of days earlier, I'd seen many trees felled in residential yards — big trees; a source of shade in this dry hot country, and a place for squirrels and birds to forage and nest. Felled, they brought to mind "Binsey Poplars," by Gerard Manley Hopkins (from his *Poems*), a few lines of which echoed in my mind for days:

> My aspens dear, whose airy cages quelled,
> Quelled or quenched in leaves the leaping sun,
> All felled, felled, are all felled...

It was hard, driving past them, not to think of sandalled feet in the heat of summer, dangling, as a child dreamed in the boughs. To think of sunlight filtering into the inner shade of the trees, the latticework of needles casting an intricate shadow. At the rest area east of Savona, I think of the lines again, with sadness. My dear pines, *whose airy cages quelled* the intense heat of a summer sun, on whose limbs darted the pine grosbeaks, the Clark's nutcrackers, down whose trunks pygmy nuthatches made their circular waltz — not all gone, not yet, but certainly diminished.

I want to look at the crafts at the Big Sky gas station at Skeetchestn, so we pull in. There are some pieces of jewellery in a glass case and a few baskets, some of birchbark with coloured quill or imbrication work, and a couple of small pine-needle baskets. They aren't as fine as some I've seen, but they give me hope. I imagine them as the work of girls learning from older women, their fingers learning about tension and scale. I'd like to ask to have a closer look at them, but the two women at work in the store are swaying to Patsy Cline, one of them holding the arms of the other, laughing about something one of them has said in a very low voice. No one notices me, examining the baskets intently; no one offers to open the case. I don't want to interrupt an intimate moment as the women fall to pieces with Patsy, so I leave.

At Red Hill, south of Ashcroft, I gather more needles, these from dead trees. A magpie swerves from the branches of one, heads over the hill. And I can hear a woodpecker knocking. At Nicola Lake, I hear a pileated woodpecker and think how rich the feeding would be for them — larvae and beetles in the millions. Yet the numbers are barely touched by birds because everywhere, pine trees are dying.

The final stop for gathering is Skihist, near Lytton. On the bench stuck out over the Thompson River, the needles — all from living trees — are the longest yet. I suspect this is because the pines here are growing in the grassy picnic site, which is watered regularly. On the Sunday morning in late September when we stop for a walk and for me to collect needles, the sprinklers must have just shut off because the grass and paved pathways are wet, a few squirrels drinking from the puddles. I've seen pine-needle baskets made by a Nlaka'pamux woman in Lytton; exquisite, tight constructions, the stitching as fine as embroidery. I wonder if she uses needles from irrigated trees; if she's found that they are suppler, more pliant. Already I'm making assumptions about something I know almost nothing about. I label the Skihist bundle, though it's not really necessary. These needles lie on top of the bucket like leggy models among ordinary girls.

Then home down the Fraser Canyon where the pines thin out and dwindle by Boston Bar, though a few are still straight-standing and tall down near Alexandria Bridge.

So many beetles are beautiful. Metallic green, bearing curved horns, limned with gold: no wonder they inspired jewellery fit for a pharaoh. The buprestids are as lovely as any, some of them iridescent blue or green, with coppery detail. Other species are lacquered black and red or yellow, small eyes painted onto their cuticle.

This summer I was sitting on our deck, reading and making notes, when a huge beetle landed on my book. Its body was at least a centimetre long with antennae twice as long again. The body was almost bronze, with silvery speckles, and there were blue spots on the elytra or forewings. I watched it for ten or fifteen minutes as it walked on the table, then flew up to the eavestrough on the roof of the sunroom adjacent to the deck, its antennae waving elegantly as it searched the air.

I left to find a field guide, and when I returned, the beetle had disappeared. I'd made a descriptive note while I watched it, though, and was able to determine it was a *Monochamus maculosus*, or spotted sawyer, a beetle perhaps as potentially dangerous in its way as the mountain pine beetle. Their life cycles are similar — the females cut into bark and lay their eggs within the tree (they attack pines, spruce, and fir, depending on which species of *Monochamus* they are, and where they live in North America), and then the larvae bore tunnels, eating soft tissue, until they pupate; the adults eat their way out of the tree.

It's not difficult to imagine an army of spotted sawyers intent on devouring a forest, their formidable antennae searching and interrogating as the insects mount likely trees. One would see them, feel them underfoot the way people have described a plague of locusts, the cracking of their shells everywhere. But I think the difference is scale. A series of mild winters has meant that the mountain pine beetles have multiplied beyond anyone's wildest estimation and although tiny in themselves — seemingly vulnerable when seen in photographs as larvae curled up in their galleries — in multitudes they have decimated huge tracts of British Columbia's forests.

Consider the numbers of beetles themselves, Biblical in proportion, gnawing their way through the Interior. In ancient Egypt, the dead wore special scarab amulets over their hearts, in part to ensure resurrection, following the belief that Khepri, a scarab god, pushed the sun from the eastern sky to the western horizon.[2] (As the dung ball created by a scarab beetle was pushed and buried in earth to hide it from predators, so the sun rolled over the horizon into the ground for the night, to be reborn in the east each morning. A perfect enactment of birth and death, of rising and setting.) In "Beetles As Religious Symbols,"

Yves Cambefort tells us that an Egyptian writer, Horapollo, in a treatise on hieroglyphs available to us only in its Greek translation, revealed that the word for "scarab" and the Greek word "monogenes" or "only begotten," used for Christ, were one and the same. (Christian authors such as John used "monogenes" for Christ in his Gospel 3:16: "For God so loved the world, that he gave his only begotten Son, that whosoever believeth in him should not perish but have everlasting life.")

On a recent trip to the Nicola Valley, we saw little white bags hanging from the healthy pines in Monck Park. Amulets, I thought — (*over their hearts*) — and in some ways, I was right. The bags held Verbenone, a synthetic version of a pheromone used by beetles to announce to others that the tree is off limits, already in use. When I did some reading about this method of repelling the beetles, the experts recommending it were as insistent about the way it needed to be used (hung as high as one could reach, on the north or shady side of the tree) as were those advising on placement of scarabs in the *Egyptian Book of the Dead*, a sign that, beyond the science, belief systems may be as dogmatic now as they were in the fourteenth century BC.

Listening: a diagnostic

I discovered the work of Alex Metcalf while reading the *Guardian* online. We used to get the international edition of the newspaper itself passed along to us and I enjoyed the nature column by Ralph Whitlock, a kind of old-fashioned ramble through the English countryside, written in a slightly formal way, full of lore and decency. I liked the book reviews, too, and this has led me to continue, in a way, my relationship with the *Guardian*: every Saturday morning I read the book pages online as well as other arts coverage.

Metcalf, a British artist and educator, has developed a device for listening to trees. The device itself is very handsome, like a streamlined ear trumpet in some sort of polished metal. What makes it interesting is that it contains a sophisticated system for amplifying sound and for filtering out ambient noise. When it's positioned on the trunk of a tree, one can hear the tree drinking. There's a *whoosh* as the xylem draws up water and nutrients, then clicking as the water is taken into cells, displacing air. *Click, click.* That's the air escaping and the water entering. People knew that trees drank before Alex Metcalf's work, but the process for hearing it was intrusive, and involved drilling into the trunk. The *Guardian* reviewed Metcalf's show at the Royal Botanical Gardens, Kew, near London, in which a number of the listening devices were hung from various trees. Someone could simply take the device, hold it to the tree, and listen. A simple but remarkable idea.

I remember hearing my unborn children's heartbeats for the first time, during an ultrasound procedure, and vowing that nothing would ever hurt them. Introducing children to the sentience of trees, letting them hear the sound of a big tree taking in water, to let them meet as organisms, one drinking, the other witnessing, could no doubt instil a deep respect which could change the way many people view the natural world.

It occurred to me that I might be able to listen to ponderosa pines drinking — well, any tree for that matter but it was pines I was thinking about at the time — and that a living tree and a dying tree would probably sound different. Or would they? I thought, for instance, that a tree dying as a result of beetle damage might be thirsty; its ability to take up sufficient water compromised by all that insect activity and the effects of the blue-staining fungus.

I tried to locate an ear trumpet, thinking that I might be able to detect drinking or lack of it (using a live tree as an example) by holding it to the trunk. I checked on eBay, at sites outfitting people involved in Civil War re-enactments (I'd discovered replica tin ear trumpets were available for those playing the roles of medics, but had no success in actually locating one). I called one son, who I thought might be willing to check antique and junk stores in the medium-sized city where he lived. He agreed to check a few places to see if one might surface.

But I had no luck finding an ear trumpet.

Then I thought, well, what about a fetoscope? Not the newer Doppler ones, but the older ones, which are essentially metal trumpets, or wooden? That made sense. (The sound of the *whoosh* as the cells took in water is not unlike the rush of blood in the placenta.) The doctor who delivered one of my babies is now a family friend. He thought he could find one to lend me. It didn't happen, though. He was away, and then there was a death in the family, and I felt shy about asking again.

I thought some more. What about a stethoscope? I called a recently retired nurse in our community, and she was very willing to lend me not one but two stethoscopes, one with better amplification than the other. There might be a difference, she said, and that in itself would be interesting. We arranged for me to collect the stethoscopes, but she forgot to leave them for me on my way to the ferry I was taking for the first leg of our journey into the Interior. I returned home to an apologetic message from the nurse, saying she hoped it wasn't too late. By then I was resigned to the fact that listening to trees, living or dying, wasn't in my immediate future.

Following simple instructions

The day after we arrive home from the Interior, I fill the kitchen sink with warm water, add a little laundry soap and two tablespoons of bleach. Beginning with the last bundle collected, I gently wash the pine needles, swishing them through the soapy water, and rinse them briefly. I wrap them in a soft towel and wash each of the other bundles, too.

The messy collection of needles from Nicola Lake has to be sorted. Broken or obviously damaged ones discarded, all of them laid out on the counter with sheaths at the top, and then washed in what I think might be suitable amounts for working with at one time. This is hard to estimate, because none of the material I've read specifies what quantity is needed for a small basket. One book sells pine needles in four-ounce increments, which should give me a rough idea. I quilted for years and have recently taken up knitting, something I did in my late teens and early twenties, but not since; I can estimate cottons and yarns reasonably accurately. But looking at needles heaped under a tree or sorted into bundles, I am lost. Even though I have a small scale, I don't weigh the bundles, thinking that I ought to figure it out by looking, imagining.

For a few days, I simply look at the washed bundles, wondering about them, reading and rereading the instructions for beginning a pine-needle basket. I have real difficulty translating the written words with their line drawings into something I can envision: my fingers handling the materials in a confident way. One version of the first step has the weaver moving from left to right, like reading or writing. The other version, considered more "authoritative," has one working from right to left. I look at my hands, wanting them to decide. They're mute, unable to help.

Take a bundle of three pine needles (or two bundles, says one book). So take your three, six, or even eight pine needles, and snip the sheath(s) off the end (or pinch off between your fingers). Make a little circle of one end with a tail of the sheath-end of the needles. Thread a length of raffia onto a wide-eyed tapestry needle and begin to wrap the circle, drawing the tapestry needle from back to front. Oh, I am all fingers, all thumbs! No dexterity at all. I study the diagram, wrap my circle, drop my pine needles, and unravel my raffia until it is impossible to work with. I begin again, burning that first attempt. Then the second.

The fire welcomes them, the dry needles and crisp raffia going up in smoke that floats above our house, providing for a moment an echo of those campfires years ago when we sat with our children under pines in the evening while loons called out on the lake and a few bats darted above us. When I tucked them into their sleeping bags and smelled smoke in their hair, the greeny water of the lake.

The third try is a little more successful. I've learned to choose a strand of raffia carefully and to keep a second needle threaded in case of accidents.

Although I study the instructions on how to make a variety of stitches — chain, open V, diamond, wheat, fern, and Indian wrap — I can only just make the simplest of stitches, something without a known name, to keep my pine needles in place. The basket I gave my husband is characterized by the most delicate stitching, something to aspire to, or else feel the weight of, against the fingers, and perhaps give up basket making completely. I'm determined to try harder.

Despite my clumsiness, the little beginning grows. I've managed to figure out how to slide new bundles of needles, their sheaths removed, into the ones that have already been stitched into place. Around, around, my fingers tuck and stitch, the needles coiling obediently. It's a kind of dance, the slow waltz of turning and holding. I gradually begin to leave more space between stitches, trusting that the coils will stay in place.

Sometimes an image falls into your hands and becomes part of your thinking. You cannot see anything else (or so it seems) without the knowledge of this image. I have always read botanical texts of one sort or another and have viewed hundreds of pictures of trees, flowers, the root systems of grass. All of them have been interesting and some of them instructive. The before and after pollination pictures of fawn lilies, for instance — the demure tepals nodding, then turned back after being fertilized by bees or other insects: a sign to say "no nectar; don't bother." I had simply thought the turned-back tepals were a sign of maturity — and in some ways that's exactly right. The language of flowers is as provocative as any other kind of body language.

The image I am thinking of now is of two pines, one of them hung with cones, from the Carrara Herbal, a parchment codex made in the very last years of the fourteenth century in Padua. I encountered the illustration in Anna Pavord's *The Naming of Names: The Search for Order in the World of Plants*. It's a marvellous book, taking the reader from Theophrastus (c.372 BC-287 BC) through to the seventeenth-century British flora of John Ray, providing a history of the development of botanical nomenclature. I was familiar with some of the plantsmen, illustrators, and herbalists who form the cast of characters in Pavord's book, but others were utterly new to me, and I confess I fell in love with the anonymous artist who illustrated Jacopo Filippino's text for Francesco Carrara, the last lord of independent Padua. This text is an Italian translation of a medical-botanical treatise, *The Book of Simple Medicine*, written in the twelfth century by an Arab physician, Serapion the Younger. The Arabic treatise is very close in content to *De materia medica* by Pedanios Dioscorides, written in Greek but known by its Latin title, compiled just as the first millennium turned.

My Italian is minimal, though I have a grammar and a dictionary and can puzzle through short passages with some understanding. However, Jacopo's translation is an archaic Paduan dialect. There is a commentary[3] on the Herbal, in contemporary Italian, and I have some of this to consult: pages photocopied by a son with access to the one copy in university library holdings in Canada (the University of Alberta).

Art historian Sarah Rozalja Kyle, who specializes in the Carrara Herbal, has been very helpful: she directed me to *De materia medica* by Dioscorides, knowing how close the two texts were, particularly the entries on pines, and knowing that *De materia medica* is more easily available in English translation. So I do have some sense of the relationship between text and image in this work. Back to the commentary on the Herbal: the entry on pines is intriguing. Like Dioscorides, Serapion (via Jacopo Filippino) tells us that the bark can be pounded finely and used in the treatment of boils. Decoctions of inner wood are used for painful teeth and uterine disorders. The wood itself seems to have been made into pessaries to hold a prolapsed uterus in place or to prevent incontinence. (How times change, and don't change! Pessaries are now made of plastic or rubber or silicone, but serve the same function. I assume that pine was used because the antibiotic qualities of the resins helped prevent infections.) I learned that pine cones are useful also for urinary problems and the corruption of the humours.

Reading this material is a window into another time, when plants were an important part of healing, a window that also acts as a hinge between what was known before that and what we know now. I mean by this that Dioscorides and Serapion took a leap forward from Theophrastus, also discussed in Pavord's book, and who impresses the reader with his keen powers of observation. He wrote many works but only two of those concerned with botany survive, *Historia plantarum* and *De causis plantarum*, written sometime around 300 BC but pretty much forgotten until translated into Latin in the thirteenth century. A Greek from Lesbos, Theophrastus worked hard on a system of classification, of plant structures, and reproduction but he didn't understand the mysteries of pollination. When he describes the catkins of filberts, you want to nudge him that little distance to realize what was happening that allowed nuts to form. (He knew that something occurred when the spathe of the male date palm was shaken over the flower of the female. But he didn't actually know what this had to do with the formation of fruit. And who can blame him? It's not as though all that people came to learn over the next twenty-three hundred years was available to him. And how many people now could readily write so attentively of filbert catkins? "The filbert after casting its fruit produces its clustering growth...several of these grow from one stalk, and some call them catkins."[4]) So I'm quite fascinated to read the texts that build on that body of knowledge handed down by Theophrastus, texts which accommodate more specific observation about cause and effect, hypothesis and proof: in short what we know as science.

But really it's the gouache portrait of the two pines that I love most in Anna Pavord's book. The artist has outlined with green pigment and then delicately painted on the foliage, even cones on the tree on the left. Grass grows delicately beneath, each tussock articulated as individual and discrete.

There are other portraits in this codex that are ravishingly beautiful. A leafy melon plant with a single opulent fruit that occupies two-thirds of a folio. A *Zizyphus jujube* with a bit of moss on its trunk. Grapevines so life-like they could be photographs, the tendrils reaching for something to attach themselves to. The unknown artist of the Carrara Herbal observed plants in their natural setting and allowed them equal ease on the page. There is such joy in his renderings, exuberance in his line and detail. And yet he never finished the work. Space was left for many more than the fifty-six portraits that exist. It seems that the fate of the patron Francesco Carrara—deposed in 1403 and strangled in prison three years later—was also the fate of this Herbal with its flowers and vines, its lively trees, its beguiling barley and asparagus: vitality cut short, as a root might be severed from a stalk or a seed head pinched off too early.

> Furthermore, the rings in the branches that have been cut off show the number of its years, and which were damper or drier according to the greater or lesser thickness of these rings. The rings also reveal the side of the world to which they are turned...
>
> — Leonardo Da Vinci, *Leonardo on Painting*

How in age a tree remembers, how the feet of tiny birds felt on the bark; how on a summer day, drowsing in sunlight, a tree might have been startled awake by a bear climbing to its first strong branch; how an osprey might have settled on the broken crown to survey the lake, the glittering run of river. How the pines stand in their wild observatories, anchored in rock, looking to the heavens, drinking deeply from the aquifer. They have seen meteorites fall, leaned into wind with sockeye migrating below them; given a small shake as ash from burning forests settled on their boughs.

How in age our own bodies remember their youth, how it felt to make love on bare ground (pollen drifting from one cone to another), to rise and walk among trees, light shimmering through their leaves. Listen! A nuthatch, a grey jay, a woodpecker, feasting on insects. How time compresses, so that all summers arrange themselves in a codex of dry skin, tart berries on the tongue, the surprise of cold water as we entered rivers. How later, organizing the photographic archive, we try to imagine ourselves back into that tent on Nicola Lake, our children racing down from the volcano, the pines filtering early morning sun so beautifully that later we say, "It was paradise."

Trace the long history of weather in their wood, the variability of rainfall, drought, solar storms. This is one way to read their rings:

$$\text{Mean sensitivity} = \frac{1}{n-1} \sum_{t=1}^{t=n-1} \left| \frac{2(x_{t+1} - x_t)}{x_{t+1} + x_t} \right|^5$$

The ring width is x, t is the year number of the ring, and n is the number of rings in the sequence. Another way: take a cross-section from that felled tree at the lookout over Kamloops Lake and run your thumb over the wood. Read invasions in the tunnels, read fungus in the blue layer, read a hundred years of trains on the opposite side of the lake, read drought in the fact of its death.

Nearby, the early texts of trees haunt the strata of the Tranquille Shale, sediments of a shallow lake from the Eocene epoch. Dawn redwoods, ginkgos, white cedars, pines, elms, alders, beeches, and others yet to be identified (no obvious modern descendant) have left their traces in the sedimentary layers. Where they grew, flowered, and died, mosquitoes hummed, something with feathers hovered, the small *Eosalmo driftwoodensis* darted in the water.

When we stopped at the fossil field this past September, thinking to join a dig, it was hot, no one was around, but then a voice called down from the cliffs, echoing in that still air like gunshot. For a moment, it was like being in another story, one set in Navajo country, red rocks glazed with heat. The trail up was very steep, and a hawk drifted. Grasshoppers clicked in the dry grass. The thought of prying fronds of extinct cedars or a tiny winged maple seed from their long sleep in shale was suddenly unthinkable...so we turned around on the shelf of rock above the highway and drove away.

Change Partners (or, The Last Waltz)
Again and again I begin a basket. The needles are brittle, even after their soaking in warm water. I make a clumsy loop, then throw it in the fire. I found the raffia too awkward to work with the first time, the way it unravelled and split, broke at critical moments. After talking to the woman who made the basket I bought my husband for Christmas two years ago, I decide to try using waxed linen thread instead. (Her basket has that delicate stitching — the tracks of tiny birds across the smooth woven surface. And I'm not sure why it took me so long to think of contacting her for advice; when I bought the basket, she tucked a business card into it with her phone number.)

I can't find the thread anywhere on the Coast. One person recommends the Internet. A friend offers to look in Vancouver; she comes back with a bag containing eight small skeins of six-ply linen, like embroidery floss. I asked for unbleached linen and she's brought some of that, as well as a pale brown, a russet, a green almost exactly the colour of sage. And it's expensive! My bags of raffia cost a dollar for each two-ounce bag. (Two ounces of raffia is a lot!) Eight skeins of six-ply linen cost slightly more than twenty dollars and even separating the lengths into three-ply strands (which is what the woman advised me to do), I can see that it won't be a case of a little going a long way. I'll have to be careful.

I melt a sheet of pure beeswax and pour it into a little tart tin. When it hardens, I have a block to run my thread through, to strengthen it. The smell is lovely. I'm taken back to the Nicola Valley apiary on the Coldwater Road where we went once to buy honey on a fall day. Walking around to the shed where the processing went on, we passed a whole lot of wooden hives at one end of a loading area (we recognized them from the road into Monck Park where hives are delivered in spring for alfalfa pollination), drowsy bees hovering, and the scent of honey so strong in the air that I stopped and closed my eyes for the sweetness.

I begin, and begin again. Once, twice, three times. Fourth time lucky. I like the way the linen thread slides smoothly through the coiled pine needles. Around and around I wind and stitch, a slow and serious waltz of materials, turning and backstitching to anchor a tail of thread; weaving in the new trio, my wide-eyed tapestry needle straight-backed and poised.

Anticipate the second beat. Hesitate a moment. Then select, and soak, needles from Heffley Creek. Introduce them into the circle. They're brittle and stiff as I coil them and wrap them, not easily coaxed into pliancy. Wrap them with Nicola Lake needles again, supple and long. The stitches are longer and smoother than they were using raffia. And the base is growing, widening, a little dance floor of sweet-scented pine and waxed fine linen.

Change partners! Introduce needles from the lookout over Kamloops Lake where a nuthatch danced along the length of fallen pine. They settle into the small space, the embrace of the Heffley Creek clusters, and I ease thread over and through to three-quarter time.

The basket grows, perhaps eight centimetres wide at its base; I've begun to build up the sides, turning and wrapping. My hands smell of resin, the drops of sap that linger on some of the clusters of pine needles, and my thumbs are rough with smoothing the coils beneath the threads. I think of waltz tunes — the strong beat followed by two lighter ones, the second one pushing into the third, the hand at the small of the back. I think of "Swan Lake" and "Moon River," "The Brand New Tennessee Waltz" ("You're literally waltzing on air"), and I hum as I stitch, hum and sway a little as I remember the trees in wind, rapturous and free, now tunnelled with beetles and fallen.

Because most of the needles I'm using are from dead trees, I look at my bundles, all cleaned and tied with string in their cardboard box, and realize that the geography of them, the narrative of their collection, represents a dying landscape, bracketed by living trees — Nicola Lake and Skihist.

The next partner in this leafy waltz is from Red Hill, a rest area south of Ashcroft on Highway 1. It's on the east side of the highway, just a few kilometres south of where the Hat Creek Road meets the highway. (That is one of the prettiest roads on earth, following Oregon Jack Creek up into the upper Hat Creek Valley where ranches are spread out on the top of the world. You can follow the Hat Creek Road up and over to Highway 99, the peaceful road to Pavilion and Lillooet. When I went into the upper Hat Creek Valley in 2003, I don't remember dying pines. But in the fall of 2006, we were shocked to see the red expanse of them as we drove out Highway 99 to meet 97 at the Hat Creek Ranch, a sight sobering enough for us to each have an extra glass of wine when we stopped for lunch that day at the Bonaparte Bench winery.)

In the past, we'd always paused at Red Hill to give our dog a chance to stretch her legs when driving north or south on Highway 1. The needles from these dead trees are the shortest of my collection; the longest bunches in this group are only fifteen to sixteen and a half centimetres. Given that the range for ponderosa needles is eighteen to twenty-three centimetres, these are clearly the runts of the litter. It's hard to imagine a drier location for pine trees than the parched benches from Cache Creek south to Spences Bridge. Arid hills, flinty rock, everything grey and pale green and tawny, an ancient wooden church at the foot of a talus. Yet the soil is obviously rich enough to support pines; and the productive orchards and gardens down towards Spences Bridge have responded eagerly to irrigation. Given water, who knows how voluptuous these lean ponderosas might have become? My fingers are now accustomed to this work, and I'm certain that the Red Hill needles are thinner than the others. But, oh, they smell sweet! It's as though they've concentrated the qualities that make them pine — intensity of scent, of slender brown leaf, of tiny beads of resin along the sheath.

The Red Hill needles are easy to place. They settle into the embrace of the Kamloops Lake trios, smooth as they circle and rise to the moment when the next bunch is introduced. A hesitation here, then here, as a needle snaps, as I run out of thread, as the short length demands that I stop and choose another trio. Pause, and step, and move with the weaving. A basket of dry leaves grows as I work.

For this is the truly amazing thing: that clusters of dry pine needles and some lengths of linen or raffia twine can grow into a basket. Who discovered this? Who sat with a pile of needles and some cordage of stripped nettle fibre or root and figured out how to bind them together in this way that is simple yet extraordinary? Who lifted her fingers to her face and breathed in the heady smell of summer, of heat and resin? Who imagined a gathering of spring beauties and Saskatoon berries inside it, waiting to be mashed into cakes?

There are baskets dating back to 4000 BC, associated with the Egyptians and burial; they contained food for those who accompanied the royal dead to the afterlife. And baskets were used to form pottery before the invention of the potter's wheel. (Stone Age pots often have the pattern of their wicker moulds on their surfaces, although of course the baskets themselves were more perishable. The pots survived with the shadow of their forms on their ancient sides.)

By 1569, there was even a Worshipful Company of Basketmakers[6] in England, part of the livery or guild tradition in the City of London. They were suppliers to butchers, among others, who would have valued the practicality of baskets for holding and draining meat and viscera, perhaps into a waiting basin so the blood could be kept for pudding; the baskets could be easily rinsed and dried afterwards. (I think of the beautiful open-work baskets made of cedar withes by the Northwest Coast peoples to drain clams and other shellfish, and how the berry baskets in museum collections almost always retain a slight stain from the juices gathered so long ago.)

The final partner on the dance card: the long needles from the Skihist picnic site, high above the Thompson River as it races to its meeting with the Fraser, entering the waltz. These are a pleasure to work with, so supple and pliant, easily circling the rising rim of the basket. My hands have taken to this in a way my body never took to dancing. As I work, I think of how a basket is more than a sum of its parts, as anything marvellous is. Some dry pine needles collected under trees, raffia, and linen thread pushed through the eye of a tapestry needle. A lump of wax in a foil tart tin. A book for instructions. A bowl to soak and rehydrate bunches from Monck Park, Heffley Creek, Kamloops Lake, Red Hill near Ashcroft, the high bench of Skihist.

When I finish the basket, I hold it in my hands with the kind of tentative wonder one holds a small bird dazed from a collision with a window. It almost breathes, this basket. What will it hold? Loose change, a necklace, a few nuts in their shells at Christmas? Already it contains a story, though the narrative is perhaps a little quiet for these troubled times. A road trip taken in the early fall, just as the aspens were turning in the Nicola Valley, with stops at Brookmere, Quilchena, the Lundbom Forestry campsite where we ate a meal with friends encountered unexpectedly as they drove in to camp and we were leaving after a walk around the quiet lake ("Why is that motorhome honking?" "I don't know. A mechanical problem, maybe? We'd better stop." "It's Solveigh and Joe!" "No!" "Yes, it is!").

A drive up towards Sun Peaks Resort where the opulent buildings sent us back again to the solace of the Knouff Lake Road and its ranches, wild-eyed cattle standing in dust as thick as snow on the side of the road. Then returning home west to Cache Creek and south along the Thompson, the Fraser, stopping in likely places for pine needles. In this story, the children have all left home but their ghosts still run down into the kikuli pits at Nicola Lake, gather pine cones to burn in the campfire after shaking them first to release the seeds, and beg for another hour of play before settling down for the night in their sleeping bags. During pauses in the telling, loons call, a coyote yips, and listen! Wind off the lake stirs the wild roses by the shore. Or is it a bear feeding on the blushing hips?

This basket is too small to hold much more than memories, though in a way the world is constructed of such things dreamed into being and remembered in all their textures: pine needles, the stray feathers of a nuthatch, a dazed bat found once under bark, emerald green beetles in flight or tiny brown ones burrowing into healthy trees and leaving as a death sentence the strange scribble of their life cycle. Remembered as a gracious dance of the living and the dead in the perfection of sunlight. As though memories are enough to feed the beloved in their afterlife, as though nothing else would do.

Fagus sylvatica

Traces

This is what I know. My father's father, ИВАН КИШКАН came to North America in 1908. He landed at Ellis Island, having left his home in Bukovina; his village was Ivankivtsi, in the Chernivtsi oblast. He was a miner. When he arrived in America, he went to work at Franklin Furnace, New Jersey, and from there moved to Glace Bay, Nova Scotia; Phoenix, British Columbia; Kananaskis, Alberta; finally settling for a time in Drumheller, Alberta. He died in Beverly, near Edmonton, in 1957.

Although I was not quite three when he died, I do have memories of him. Each summer my family would travel to Beverly to visit my father's parents. I remember a small house with a metal roof. During storms, my brothers and I were frightened at the sound of hail drumming on the roof. Our grandfather spoke heavily accented English, which we also found frightening, though he was not a mean man. Our father told us his father couldn't read or write but I think he meant English because there's a postcard among my father's papers, written from someone called Sam Fedoruk to my grandfather in a Slavic language. The message must have been important enough for him to keep. My father taught his father to sign his name for legal purposes. ИВАН КИШКАН became John Kishkan.

When I was very young, I wasn't interested in knowing more about him. My father told us next to nothing. "Did your father have brothers or sisters?" we wondered once. "I never asked," he replied. With a drink or two under his belt, he'd become maudlin, full of pity for his dad, implying that life had not treated his father fairly. My mother concurred. There was one little story which they repeated: when the First World War broke out, my grandfather was working in a mine in Kananaskis and was fired immediately without pay. This was because of his Austro-Hungarian citizenship, we were told. (We have a photograph of him in the uniform of the KUK Freiherr von Reinländer regiment taken during his army service from 1901 to 1904; he'd been a conscript in the land force of the Austro-Hungarian monarchy.)

Later, I learned that many men of his background were interned in the Rockies as "enemy aliens." I should have known this all along, but it's not the history we were taught at school, and our family didn't discuss it at home. The story that was told had him walking home — though where that was, exactly, is unknown. "That poor old man," my mother would echo as the story was told any time we asked about our grandfather. Yet he was just thirty-five when the war broke out — hardly old, though probably battered by life as a miner in remote camps or towns. He hadn't yet met my grandmother, a widow with eight children. Later they would marry and have Julia, who died as a toddler, and my father, who lived.

I knew almost nothing, and my father was dying. Most of his memory was gone, or unreliable. I visited him and stood by his bed while he fumbled for words or dozed off, his mouth open, mid-sentence. He had been hospitalized for urinary problems related to prostate cancer, and during the treatment process, other tumours were discovered. A week after he was admitted, he had a stroke while sleeping, so one arm was useless — he couldn't grip with his hand or move the arm on his own. A nurse kept asking him if he could pick up his cup and he looked confused, then tried to lift it with his left hand which trembled and had no strength. My mother asked him questions about his food and about sleeping and he blinked, then said, "Noo...ooo," in a perplexed but gentle way, at odds with the man I'd always known.

I told him about the herd of elk that had discovered our fruit trees and he smiled very faintly. When one of his old work mates showed up to see him, he was silent at first, then said in a weak voice, "They almost lost me last night." "Oh, they did not," my mother corrected, crisply. Understandably, he was depressed.

As my father approached death, there were questions. "How do you feel, can you hear me, are you warm enough? (Can you remember my name?)" There was remorse, sorrow, and anger. There was an impending sense of absence. Or multiple absences—because as my father was leaving, so went any knowledge of his father, all his half-sisters and brothers, his mother. That strand of my family was disappearing without much of a trace, and those tiny traces that do remain are mysterious beyond words. The photograph of the beautiful woman in my grandfather's papers—of course I wonder who she is. (I see my broad shoulders, my cheekbones.) She might be his mother or his sister but I will probably never know for certain. I imagine myself, photograph in hand, walking the streets of Ivankivtsi, asking if anyone recognizes her. Recognizes me by my shoulders, her dark hair.

He was dying as I wrote this, his own father a name on a colour copy of a passport, two photographs, army papers in a language I can't read.

The European beech, of which the copper beech is a subspecies, ranges from Scandinavia to France, southern England, the northern Iberian Peninsula, and east to northwest Turkey. In my grandfather's native Bukovina, on the slopes of the Carpathian Mountains, there are vast beech forests, intermixed with spruce and fir, holm oak and pine. Tracing the root of the place name, I find that there are versions in German, Romanian, Ukrainian, and Polish, all nations with footholds in Bukovina, past and present.

There's debate about when the beech arrived in England—before or after the ice age; before or after Caesar. European farmers have long known the value of beechnuts, or mast, as food for pigs, not unlike those swine in the *Odyssey*, "rooting for acorns to their hearts' content, / drinking the dark still water. Boarflesh grows / pink and fat on that rich diet…"[1] In times of hunger, human beings have used both acorns and beech mast too, grinding the nuts to flour and leaching out the bitter tannins by soaking the meal. Well-ripened mast was also pressed for its oil, similar to olive or hazel oil; this was used like butter in Silesia. The cakes left after the oil was pressed from the mast were fed to livestock.

When I married John in 1979, we lived in North Vancouver; there was an enormous copper beech (*Fagus sylvatica*, spp.) at the top of our driveway. We lived in an old house, built just before the First World War, and the tree could easily have been that old as well. All summer its lovely cool foliage provided spreading deep colour to drive into as we came down the hill to our house—it was the last house at the foot of a cul-de-sac; where the road ended, a ravine continued. Our sheepdog, Friday, used to sleep in the shade of the copper beech.

I suspect the copper beech near that earlier home of ours was planted by someone who'd known the tree in Europe and for whom beeches constituted arboreal perfection. In fall, our beech tree turned a glorious gold. Among the native cedars and firs, it was like an elegant visitor, richly dressed. Everything about it was dramatic: its size; the colour contrast in its leaves, season to season; its bare architecture in winter. I wondered occasionally how it had arrived on our block—what stories were contained in its smooth bark, its trunk which several generations of children climbed to get to the spreading branches overhanging the ravine. That ravine served as a wild and secretive corridor for bears and deer that came down from Fromme Mountain to feast in the gardens and orchards of the neighbourhood — animals who became elements in our own stories. How we lost pears one autumn to a bear, and how deer could be seen on summer mornings, reaching for cherries.

We bought a printing press when we lived in that house, dismantling a basement wall so the press could be moved inside. It took most of a day to drag it along temporary skids, over the grass and into the basement using a come-along winch. The press was a platen press, a beautiful late-nineteenth-century machine, with complex gears and wheels. We loved the notion of printing broadsheets and eventually small books. The press came with some wooden type, big clumsy capital letters, and we used them for a Christmas card that first year — JOY, in green, with a big exclamation mark.

In his book *Beechcombings*, Richard Mabey tells us that the origins of the word "beech" echo the German and Dutch words for book; this was in part because early European texts were inscribed on the wood and bark of beech trees. Before Gutenberg printed his forty-two-line Latin Bible, set in moveable metal type, in the fifteenth century, block books were printed in a process called xylography; text and illustrations were cut as a mirror image into sawn wood, usually beech, an echo of *bok* and *buche*, those early texts inscribed in the bark of beeches. And maybe those wooden letters we used to proclaim our joy were in fact beech. They'd need to be hardwood to withstand the great pressure of the platen meeting the type bed.

In those years, I could have asked my father more about his own father. In the newness of my relationship with the man who'd just become my husband, I was softer than the young woman who'd argued with her father, who'd carried the sting of his comments and judgements in her heart like small bleeding wounds. He liked the sound of his own voice, which precluded generous discussion. And I was often too quick to treat him impatiently, dismissively. But he had a history and part of it was mine, or needed to be in the years that would follow when my own children were born and they asked questions about the past.

My parents visited us often in North Vancouver, younger than I am now, and I should have anticipated the change that would come, like the turning of the leaves on that tree. That stories could be hoarded, like beech mast, for the time when they might be needed. And like beech mast, the hardness of the stories could be softened over time, bitterness leached away, made palatable by forgiveness and love.

Phoenix Cemetery: Lat: 49° 06' 05"N, Lon: 118° 35' 06"W
In 1911, my grandfather was living in Phoenix, British Columbia, a copper mining town founded in the mid-1890s. Mining claims were staked, cabins erected, and the CPR and Great Northern Railways arrived in 1896 to facilitate removal of the copper ore to the smelter at nearby Greenwood. In the early days of the twentieth century, the town boasted twenty hotels and saloons, four churches, a town hall, an opera house, its own telephone exchange and electric power plant, and a hockey team, which won the provincial championship in 1911. (This team requested the opportunity to play for the Stanley Cup that year, but apparently it was decided that they'd applied too late.)

This is what I know: almost nothing. When I look at photographs of Phoenix, I wonder exactly what my grandfather did. He had worked in mines ever since arriving in North America, but I really have no idea what this involved. One in ten men in Phoenix was a "timberman," constructing the prop posts that supported the excavations. Machines had to be serviced and maintained, pumps operated. I scrutinize photographs from Phoenix to try to understand my grandfather's work. It looks so dark and precarious in the shafts, in black and white. I can see picks and shovels leaning against the walls or gripped in the hands of a miner in overalls. And reading about these mines, I come across accounts of men killed when ore buckets fell on their heads or when the foul air from powder blasts suffocated them. There were fires and derailments and shootings. There were diseases.

Years ago, we drove through the Boundary country, where Phoenix had been, on our way back from a family camping trip to the Royal Tyrrell Museum of Palaeontology in Drumheller. There had been traces of my grandfather on that trip — we stopped at the site of the coal mine near Drumheller where he'd worked, and which my father said was more or less adjacent to his family's small holding. I wondered briefly about happiness — the hills were so bleak and colourless, empty; or at least that's what I saw; though my father had mentioned long walks in search of dinosaur bones, rock outcroppings rich with fossils — but I was so busy keeping my small children fed and occupied that I didn't linger much on family connections to Drumheller.

The summer of that camping trip was very hot, and I know from my father's accounts of his childhood that the winters were perishingly cold. I had found the Drumheller landscape bleak, without the luxury of trees and verdant valleys, though I suspect if I went again, I'd see beauty of another sort.

On a late September day in 2009, my husband and I drove from our motel in Osoyoos to Greenwood. The Boundary country is beautiful, with rolling pastures fringed with pines, aspens turning as we drove farther east. Greenwood was idyllic on the morning we arrived. We visited the museum, after espresso and pastries in the Copper Eagle Cappuccino and Bakery; we took ourselves on the walking tour to see the lovingly restored houses and buildings, all with their modern gloss of appearance in *Snow Falling on Cedars*, a movie set on the San Juan Islands but filmed in part in Greenwood and now given a place of pride in the town's history. A sign on the side of one brick building, cleverly faded to suggest a long presence, is a remnant of the film; it advertised San Juan Island strawberries. This seemed slightly ironic to me, though I'm sure the infusion of money was welcome, and the legacy of that moment of fame lingers on.

When I asked about Phoenix in the wonderful Greenwood Museum, the woman at the desk gave me a map for a self-guided tour of the Phoenix Interpretive Forest. She said there was not much left up there. We walked briefly over the site of the old smelter at the Lotzkar Memorial Park, a moonscape of dark heaps of black glass, and bell-shaped slag, and decided to drive up to Phoenix anyway, thinking that we could continue on to Grand Forks on the other side of the mountain.

The self-guided tour was organized to direct a driver from Grand Forks to Greenwood over Phoenix Mountain so we were, in essence, doing the tour backwards. This meant rapid little math figurings to determine where we were, kilometre-wise. But I rose to it, subtracting the small sums, and we paused at the switchback—our first, the guide's third, at 20.5 on the Phoenix Interpretive Forest Road, which turns off 19.2 kilometres from Grand Forks, or 0.9 from our beginning—to see the view of Greenwood and the slag pile. Then the Forshaw Homestead, where a pair of dogs barked like crazy and raced along the inside of their fence as we stopped to take a few photographs of the old farmhouse. The Coordinated Resource Management Plan Sign didn't seem worth stopping for, but the Phoenix Cemetery certainly was.

The first grave we encountered was for a woman who was buried with her infant twin sons. A pair of teddy bears and some plastic flowers graced the top of this grave. I stood in the late September air, cool but bright, and tried to imagine how the husband and father might have continued after this loss, coming home after work to empty rooms.

In that stillness, a grey jay caught sight of us and swooped to a nearby tree, eager with gossip. But nothing he said provided me with more than a fugitive understanding of the place and its stories. Men, mostly; some women; several graves with resting lambs and inscriptions indicating a nine-day-old baby or very young child. A child and a father within days of each other in 1918 when Spanish flu killed millions around the world. Some Masons. Some members of the Independent Order of Oddfellows. Men from Sweden, Finland, England, Wales, Italy. Clusters of deaths which indicated mine accidents of some severity: July 5, 1914.

Almost all the pickets had been freshly painted, and there was evidence that the work was ongoing: a brush and can tucked into a sheltered area, drips of white paint on the grass. And such dignity in the inscriptions: for a baby, two days old when he died, "Budded on earth, to bloom on heaven"; for a man from Sweden, twenty-eight years old, "Here rests a woodman of the world." There were also a number of wooden tablets, weathered and worn, with names carved in, almost gone. These wooden monuments were poignant — for their economy and for the way they are clearly modelled on expensive stone tablets. They are shaped as carefully, some of them erected (the inscriptions tell us) by family members from across the continent, or the world. Rusty tin cans hold a few wild flowers or plastic roses, and it's obvious people come to replace the flowers and prop up falling monuments.

The jay inclined its head as we walked slowly through the cemetery, touching the wood, leaning closer to read the fading names and dates. He'd seen this before, and was hoping for food.

I made a gathering of names of people who died the year my grandfather's name showed up on the census:

> WILLIAMS, THOMAS H, b. North Wales, d. Dec 07, 1911, age: 40yr
> SHEA, EUGENE P, b. Jan 01, 1869 Saranac Lake NY, d. Dec 03, 1911
> MICHELA, ANTONIO, b. Feb 10, 1873, d. Jun 13, 1911, born Aglie Corino
> MARTIN, GUSTAA ADOLF, b. Sep 14, 1886, d. Feb 02, 1911
> EVANS, JOHN, b. North Wales, d. Mar 20, 1911, age 32yr
> COOK, NANNIE A, b. May 06, 1880, d. Oct 09, 1911

I wonder if any of these were his friends or fellow boarders in the place where he was living. I can't find them on the census. Most of the names on the list have "Boarder" beside them in thick black ink but small clusters with the same name seem to indicate cooks, children, a foreman.

I wonder if my grandfather knew children in the boarding house, and if any of them might still be alive, with a vague memory of a stocky man with a strong accent. Or if he worked with young boys who might remember him still, or their sons, the ancient men you find in small towns like Greenwood or Merritt, their eyes still bright and curious, and eager to tell you what they knew in their youth.

In Alice Glanville's *Schools of the Boundary: 1891-1991*, I came across a telling moment in the School Inspector's report for the Phoenix School, 1905: "For the past year the school had an enrolment of 137 students and an average daily attendance of only half that number. A curfew by-law is urgently needed."[2] Were they truant, those absent children, or were they working, their small bodies underground with the horses and picks?

I was surprised, during the rest of the drive, that there really wasn't much left of that bustling city. The books and articles I have about Phoenix describe a true place, and it was hard to believe it had disappeared.

The first log cabin was constructed in 1895 and a photograph shows it, plain as anything, roof planks cut any old way, but the men lounging against it — including developer, promoter, and future mayor of Phoenix, George Rumberger — suggest that big plans are being made. Several men are in suits; one wears a rather formal derby hat. Confidence is in the air. A photograph taken just a year later shows a CPR terminal, several shaft houses, a crushing plant, bunkhouses, stores, quite a large hotel, a couple of trim houses. By 1901, there was a hospital designed by Francis Rattenbury, and a three-storey miners' union hall with a banquet room and theatre. When the town was abandoned, the houses and buildings stood for some years, overseen by a man called Forepaw for the hook he had instead of a hand. His real name was Adolph Sercu (or Cirque) and he'd come from Belgium to work in the Boundary mines. He was hired as caretaker of the townsite, which he accomplished from his base in the city hall where he'd moved after cutting out a star from a soup tin to wear as a badge of office.

The buildings crumbled or were scavenged for materials. In 1927, the hospital still looked intact, though it was ruined inside. When the Granby Company began an open-pit operation at Phoenix in 1956, what was left of the town was bulldozed into oblivion, apart from the cenotaph that was erected in 1919 with the proceeds of the sale of the covered skating rink, where hockey games had provided such entertainment in the town's heyday. The cenotaph had been moved from the townsite to its current placement on Phoenix Road. We stopped to look at it. It was inscribed with a line from Horace's *Odes* (Book III:ii): "*Dulce Et Decorum Est Pro Patria Mori.*"

Sad, to realize how things change and don't change—that young men still rush to war and that the resulting monuments still offer such lies to their memory, though, "It is a sweet and honourable thing to die for one's country" had perhaps not yet publicly acquired its harsh and ironic gloss in 1919. Wilfred Owen's poem which uses part of the line as its title, "*Dulce Et Decorum Est,*" wasn't published until 1920, after Owen's death. Its fierce admonishment wasn't part of the general discourse:

> If you could hear, at every jolt, the blood
> Come gargling from the froth-corrupted lungs,
> Obscene as cancer, bitter as the cud
> Of vile, incurable sores on innocent tongues, —
> My friend, you would not tell with such high zest
> To children ardent for some desperate glory,
> The old Lie: *Dulce et decorum est*
> *Pro patria mori.*[3]

And yet these young men deserve their monument—to bravery, to courage (think of them setting out from remote Phoenix in its valley of fireweed that summer of 1914, the horrors in their immediate future unimaginable), to the loss of everything they might have contributed to the world. Some of them had no doubt skated in the rink sold to finance their monument. We photographed the cenotaph in its wild loneliness, visited by people like us on days when the road wasn't mired in mud or heavy with snow.

When we came down off Phoenix Mountain to where the road meets Highway 3, I was filled with emotion but it was mysterious or at least baffling to me. I'd hoped for something of my grandfather, though I knew it was unlikely. Where he had lived in 1911 had been bulldozed under, covered with tailing dams, a lake created for Phoenix's water source and now home to rainbow trout and sunfish. We didn't venture into any of the underground workings, scared away by the signs warning of danger. I've seen the photographs, though, of these ghostly caverns with their open stopings, a man dwarfed to one side. I tried to make a connection with my grandfather but felt only the abstract sorrow I'd experienced in the cemetery. The cemetery, though, was perhaps the place where he might have stood, on ground I stood on almost a hundred years later, his hands folded, at the funeral of a friend, under small trees now grown to full and dignified height, while an ancestor of the grey jay hovered and scolded.

So far from Bukovina. So far from Drumheller, where my grandfather met and married my grandmother; from Beverly, where they moved during the Depression to that tin-roofed house where my brothers and I cried in fear during a hail storm. From Cyrillic to English; from one name to another; from the single status listed on the census to fatherhood (and grief as his first young daughter died of diphtheria at three years of age in 1924). From the beeches of his birth country to the lodgepole pines, larches, and aspens of Phoenix Mountain — though the townsite was stripped of its trees in those heady days of its origins, when he walked out in early morning to the shaft.

My grandfather was accustomed to vegetable gardens; he came from peasant stock that grew what they could or risked hunger. Probably they were hungry anyway, which is why my grandfather came to North America and sent money to bring a cousin over a few years later. Did he expect gardens in Phoenix? The short growing season meant lawns and flower borders were scarce, though lilacs flourished, and when we stopped on the roadside for our lunch, I was astonished at the size of the wild strawberry plants. I've read that potatoes were grown, and that lettuce did well, though a man working long hours digging ore or filling carts with it, or whatever it was my grandfather did, needed something more substantial than lettuce. Still, he would have known meals of potatoes and perhaps not much else in Bukovina. One of only a small handful of stories I have about my grandparents is about them digging potatoes in Drumheller in October 1926. My grandmother realized she was in labour and prepared to leave the garden. My grandfather asked, "Aren't you going to finish your row?"

For years my father remembered particularly delicious potatoes, and his mother's butter, and the foods of her homeland — cabbage rolls, perogi, noodles made with eggs from their own chickens. The butter and noodles were often sold or bartered for essentials during those bleak years of the 1930s.

In Royal Jubilee Hospital, someone had to feed my father because his right side was paralyzed.

Two of my brothers travelled to Victoria to visit my father and to help my mother. My third brother was estranged and wrote sad emails to say that he wanted no part of it. I thought of the immediacy of communication, that an email is transmitted to my electronic mailbox within seconds of being sent, a phone call connects two parties at the touch of a button, even a letter travelling between two cities takes three days at the best of times. We fly at the drop of a hat — or a message to say that a father has been taken to hospital and not expected to live. We think nothing (or everything) of a holiday in Mexico or Paris.

At the best of times, we were a family of six, with grandparents to visit in summers. But at the time of my father's last illness, spread out over Canada, with children of our own (and one brother in turn a grandfather seven times over), we had largely lost touch with one another, apart from the paperwork created by my father's imminent demise.

For my grandfather, a postcard must have been an occasion — because he kept one, a scene from the Granby Mine at Phoenix, among his papers. I know my father telephoned Beverly once or twice a year. Perhaps he wrote letters home. I don't remember. But I do remember parcels at Christmas, a cheque sent to buy a rocking chair for my grandmother who was then living with a daughter in Edmonton. I know nothing of the passage of letters or cards or cables from North America to Bukovina. I don't even know to whom such missives might have been sent. The broad-shouldered woman? The two girls standing side by side? Were they his sisters, maybe? Did he leave a mother behind, or his father? Any trace at all?

During my last visit to my father, when I was staying with my mother, my grandfather's original travel papers to North America came to light in an old envelope in my father's drawer. "Who will want these?" my mother asked, and I gladly brought the small packet home with me. Not quite a passport, but a booklet, in German, which a friend helped me to read. He was average height. Brown hair, blue eyes. We puzzled through a few of the other terms. The script was Gothic, and difficult. *Keine mitreisende* meant that no one else was travelling with him, and *Eine fahr* meant that he only had permission for a one-way journey.

While in Victoria visiting my dad, I had to walk a long corridor between hospital wings. Windows along one side of the corridor looked out to the chapel, built in 1909 — a handsome building with a five-sided apse and a stained glass window. It was nice to pass the building, and it didn't occur to me that one might enter it during a quiet moment. Then it did occur to me, and I found the entrance from the second floor of the corridor. My dad was on the third floor, so the entrance hadn't presented itself during those first few walks from the elevator to his ward.

I'm not a Christian. I was raised in a church-going family, but the concept of one god never made sense to me, especially not after I'd looked through a book about the Third Reich when I was ten and saw photographs taken at the liberation of Belsen. Around the same time, a Sunday School teacher, a woman who favoured twin-sets and plaid skirts, was fond of telling us that God loved us so much he could number the hairs on our heads (a fairly pointless activity, in my estimation). "But didn't he love those children and their parents enough to spare them the concentration camps?" I asked. She clearly wasn't prepared to answer questions like that. God seemed like a grand deception that grown-ups practised on children, and I thought that even if there was a god, he clearly didn't have his priorities in order.

The hospital chapel was peaceful, though, and I even wrote a small prayer, or more like a wish, into the book on the altar. *Let his passing be painless and graceful,* I wrote. I put his initials and my own.

Leaving the chapel, I was surprised to see a small building I recognized from *Exploring Victoria's Architecture*. I knew at once it was the Pemberton Memorial Operating Theatre, built in 1896. Large windows provided light for surgical operations, which were performed in the middle of the room. It's built of brick, with a hipped roof, and topped with a lantern ventilator. The lines are very gracious, somehow, speaking of earlier days, and I was delighted to learn that it was currently being restored and will remain a small historical presence near the chapel, in the midst of a very busy city hospital.

For me, the operating theatre was a reminder of several things — the changing face of the city where I'd spent most of my childhood, most specifically its scale (this diminutive building in the shadow of the huge hospital complex surrounding it); the idea that the past is always available to us though often not in the form that we expect it to be; that principles of design have changed to reflect population shifts (towers to accommodate the services a hospital is required to provide from childbirth to the laying out of corpses).

Being at the Royal Jubilee over those days reminded me that I'd been born in that hospital, though not of course in the Pemberton Memorial Operating Theatre. In the hospital's gift shop, I saw a photograph of the operating theatre in use in 1906. Four nurses and a doctor (and I think I'm correct in assigning these roles based on gender, given the time and the nature of those occupations) lean over a sheet-covered patient in the centre of the room. The generous windows are uncurtained, and a few flimsy electrical lighting fixtures hang from the ceiling, one of them positioned over the patient. Basins and large enamel jugs covered with cloth wait on an adjacent table while one nurse holds a jug at the ready where the doctor is operating on the patient. The table behind the team has an array of instruments on cloth; everything orderly and bright. All of this attests to the theatre's commitment to British physician and medical activist Dr. Joseph Lister's belief in the importance of antisepsis in surgical procedures; a room attached on the north side was intended for sterilization. The operating theatre's main proponent, Dr. John Chapman Davie Jr., was a keen advocate of Dr. Lister's methodologies and insisted on a space that would facilitate their practical application.

It was after leaving my father on one of my final visits that my husband and I drove off on the little trip that took us through Osoyoos and Greenwood. I had so many disparate and confusing emotions; among them the sense that we were driving deep into the interior of my relationship with my father, and his with his father, and that coloured the next few days. What I wanted to know was: what shaped my father? Who shaped him?

There is one photograph of my father with his parents. For years, it sat on a shelf in an unused room in the basement of my parents' home in Victoria. Now that the house is sold and they live in an apartment, I wonder where the photograph ended up. Perhaps in one of the three storage lockers they have for the stuff they couldn't bring themselves to sort or give away when they moved. I remember asking at one point if I could take the photograph home with me. I imagined cleaning the grimy wooden frame, polishing the glass, and hanging it in my study, though the group is so grim that the prospect of them watching me work at my desk is a little off-putting. In any case, my request was greeted with startled consternation: "That's one of the only pictures I have of my old parents."

"I just thought that as it was put away in that downstairs room that you might be willing to pass it along to me. It really needs a good cleaning and I'd like to take it home and hang it in my study."

"It's not yours. It's mine. They're *my* parents. When I want you to have it, I'll let you know."

In the photograph, my grandparents bracket my father, who must be standing on a stool because his head is at the same level as theirs. My father looks to be about three years of age. My grandmother was born in 1883, and my father in 1926, so that makes her about forty-three. She had given birth to at least ten children, two of whom died in early childhood — and she had buried one husband. I say "at least" because although I'd been awake in the night counting the children I remembered on my fingers — and coming up with eight who lived into adulthood and the two who died — there might be one or two I've forgotten. We weren't close to them, although most summers we drove to

Alberta and participated in a large celebratory meal or gathering at the lake cottage of the more prosperous family members.

The photograph of my father and his parents is hand-coloured, and poorly at that. No one is actually smiling, though my grandmother might have the tiny beginnings of one. My father is looking slightly to one side, towards his mother. He's wearing the same shirt or suit he's wearing in another photograph, equally neglected, but that one is truly strange.

In the second photograph, my father stands in his white shirt, short pants, dark stockings, and boots on a rattan chair. Someone has told him to stand still, because there is nothing natural about his pose. But—and here's the bizarre thing—hovering in the air, as though balanced on the arm of the chair, is the swaddled form of his sister Julia, who died three years before he was born. This is the late 1920s, before Photoshop—before any of the techniques we are now so accustomed to using. I know that photographers could manipulate images even in the nineteenth century (I think of Hannah Maynard in Victoria with her trick portraits and artistic interpretations). But this is clearly the work of someone who didn't have much skill at all. The half of the photograph into which baby Julia has been inserted is blurry.

That only this one photograph survives suggests that although money was probably in short supply, my grandparents wanted a record of the two children they had conceived together. Perhaps they were more sentimental than I've been led to believe, because what other reason would result in an image of a baby being inserted into the photograph of her brother-to-be, at least five years after her death? Julia was nearly three when she died, and yet the photograph is of an infant, wrapped in a blanket, wearing a hat against the cold.

Photographs are intriguing but ultimately unsatisfying. I've tried to read these ones for hidden narratives of love and family connection and perhaps I've interpreted them completely incorrectly. Still, sometimes photographs with their cryptic stories and forgotten conclusions are all we have.

In Osoyoos, I took advantage of the hotel's wireless Internet connection (at home we just have access to very slow dial-up) to find Web sites devoted to Bukovina. I loved the YouTube offering of Orthodox monks chanting psalms in Putna, a beautiful painted monastery that I know about from a book, *Sweet Bucovina*, by Ion Miclea, given my son years earlier by a Romanian classmate of his. There was also footage of the monks painting ikons, walking in forests, involved in daily monastic activities.

Having just come from a few days with my dad, and the Pemberton Memorial Operating Theatre, I was haunted by a certain familiarity in some of the chapels and buildings of the monasteries. At Putna, for example, the small tower seems to have the same hexagonal shape as the Pemberton operating theatre. The churches of Humor and Voronet — these buildings have reverberations, though of course they come from different cultures, continents, languages, and a different religious tradition. Maybe I wanted the solace of spirituality, in whatever shape it came — medical, Orthodox, polished wood, hymnals tucked behind pews. I heard the faintest echoes of liturgy, of buildings on opposite sides of the world, and they form a coherence in my heart, if not my mind.

The Pemberton whose estate funded the operating theatre was Joseph Despard Pemberton, a surveyor who was instrumental in laying out townsites in nineteenth-century British Columbia. Born in Ireland, he came to work for the Hudson's Bay Company and was involved in colonial, then provincial politics in Victoria, before founding a prominent real estate company prior to his death in 1893. His widow donated the funding for the chapel where I'd left a prayer or wish for my father's peaceful passing. When I returned to the chapel, I heard those Orthodox monks from YouTube as though they were in the room, their psalms available in the strange way that music sometimes is, across centuries, through time zones, generations, the airwaves holding and cradling the notes until they are released to the needy — believers *in extremis*, or women whose fathers are dying.

On my next visit to my father in Victoria, two weeks after the trip to Osoyoos, walking the corridor to 3C, I looked out to see the operating theatre stripped of its paint, the old brick and mortar exposed. The boards had been removed from the windows, which were elegant. The roof had been cleaned and restored. It was so beautiful that I stopped in my tracks, reminding my daughter to lend me her camera next day so I could photograph the building in its antique glory.

But alas, I forgot to bring the camera the next morning, and already workers were painting the rich rough brick a dark red. I know it's probably entirely justifiable in terms of preservation, but the plain brick was much lovelier, and I was disappointed I had missed my chance to make my own record of that beauty.

In the material I've read about the painted monasteries, like Putna (the one the monks were chanting in and which appeared in *Sweet Bucovina*), there are duelling opinions about the restoration of the frescoes—whether there is more integrity in letting them fade with time or in refreshing them to keep their saints and animals vital. In the YouTube clip I watched, I saw Orthodox monks painting frescoes and relief-carving as part of an ongoing tradition of work and worship. The frescoes are alive and lively—there is such animation in the painted walls at Sucevita, for example, where angels battle demons for the souls of those ascending to heaven. Yet here I am, wishing that the operating theatre could retain its plain brick facade. I realize I'm erratic in my expectations of this building in which I've seen echoes of the monasteries of my grandfather's homeland. I wonder if my grandfather knew about, or visited, any of the Bukovina monasteries. Were their frescoes familiar to him? Did he recognize the saints of his church in the bold figures on horses, haloed in gold? And what about those deep Voronet-blue skies with the quirky stars?

My son painted a floor cloth for me, taking a figure from *Sweet Bucovina*—an elongated horse and rider, elegant spear, the blue sky pierced with golden stars. And it spoke to me in imagery I could almost understand, the way my hand reaches to make the sign of the cross when I enter a church, even though I was never confirmed, and only ever attended Catholic services a couple of times in my life, when my father wanted company at Christmas or Easter. The rest of the year we went to a United Church. During this time, thinking about my grandfather, my father, and looking at photographs in a book about Bukovina where I find their faces (and my own) in those pages, I decide to raise the cloth from the floor to the wall of my house, my own fragment of fresco, my own saint.

I am thinking of the ways in which we are imprinted by our history, an abstract tattoo. How a small package of papers transfers names, images, a modest sum of money cabled from Franklin Furnace to Galicia in 1910, and begins the process of creating a text of my grandfather — large empty spaces scribbled here and there with a word, a booklet of military service, an image of a woman, nameless and beautiful. Even a small card, registering one Remington 1900 12-gauge single action shotgun, serial number 92005, on August 21, 1940: my grandfather's Racial Origin is noted as "Austra" (*sic*). The registration is certified by Thomas Johnson, the Chief Constable of Beverly, Alberta. My grandfather has signed in an uncertain hand, as though a child in the very beginning years of school, who has learned enough cursive lettering to form his own name. I use my finger to trace his name as though I am touching his skin, the shape of his life.

There's a story I've read in several sources, in differing forms, about the eureka moment which led to Gutenberg's first experiments with printing. As was the practice in Europe in the early fifteenth century, he had incised some letters into a piece of beech bark. Wrapping the damp bark in a piece of paper to take home, he was delighted to discover later that the text was imprinted on the paper. This sounds apocryphal to me. Paper would not have been a common commodity in those years, and I doubt it would have been available for such casual purposes, though papyrus was used and reused until about the eighth century; old scrolls were recycled in a number of intriguing ways, including as the cartonnage or wrapping around mummies. Unwrapping them, archaeologists find poems, census records, religious tracts, maps of ancient cities.

Surely Gutenberg had some experience printing block books from sawn wood. From the pivotal moment in history when he developed his skills to the point where he was printing from type, the traditional scriptoriums and copyist workshops could barely keep up with the demand for texts. Religious, philosophical, and scientific works jostled with the more practical materials needed by merchants; universities needed multiple copies of Greek and Roman manuscripts. Knowledge and information were filtering into a broader spectrum of the population, and printing facilitated this process.

Gutenberg's gift was really to bring together a number of processes which had arguably been around for a time — moveable type (the Chinese had a version in ceramic dating from the eleventh century; the Koreans used bronze a little later), using an oil-based ink, adapting a wine or olive press to apply consistent pressure from the inked type to paper, using a mixture of lead, tin, and antimony for his type, and figuring out matrices which would produce the type in uniform sizes and forms. The woodblocks of the period were carved from beech, an homage to those ur-texts on bark. Thin slices of beech were commonly used for the boards, which bound the quartos together, covered in leather or paper or linen.

My husband labours in our print shop over type, chases, ornaments, and the unwieldy nature of ink. There are far more convenient ways to transfer texts to paper, but this suits his meditative nature, and mine too, for I love to think of the slow work of poetry finding its way to a broadsheet. Paper impressed with ink, like a kiss, a tattoo. Or the history of a man I barely knew available to me in cryptic moments on paper, some of it in a script I can't even read.

He was in Calgary all day, my mother said. In Calgary. Not in the Royal Jubilee Hospital on 3C in Victoria, not home with my mum, nor fishing in one of the lakes he loved. Not even sleeping the peaceful sleep of a happy man who had no memory. No, in his mind he was in Calgary where my brother lives, the one who gave up on him, wouldn't call or write, even then, at the end of my father's life.

For some months my husband and I had a trip planned, and although part of me felt that we should cancel it and stay close to home for the duration of my father's final days, we decided to travel. My mother urged us to go, saying that there was no way of knowing how long my father would live and that I had helped her to prepare for what was to come (we'd gone to offices for forms, met with medical social workers). I promised to phone her every few days.

In Paris, I had a sense of how my grandfather must have felt stepping ashore at Ellis Island—the voices unintelligble, apart from a word here (*merci*) and there (*pardon*). On the one hand, our language is the thing most familiar to us after our families; on the other, it can create a barrier between us and the world, an invisible but impenetrable wall. My small French (*un petit peu*) as I shopped in the Marais for the makings of our dinner at the tiny flat on Rue Aubriot; his Ukrainian as his lungs were checked, his teeth examined. How brave they were, those people who walked into North America from the old forests and fields of Eastern Europe, a photograph kept in their luggage for safe passage. How willing to go underground to labour at coal seams, to sleep in railroad camps, sending small sums of money back to their families. To harness themselves to ploughs, to settle in sod houses, earth dropping on their tables while smoke from crozzling fires filled their lungs. To learn a new language. To shape unfamiliar letters into a semblance of their names.

My analogy—my French, his English—is of course indulgent. I travel in relative luxury, by plane and train, not on foot, without money, or perhaps hungry. But to find oneself a foreigner in a strange land: I try to imagine him walking from Kananaskis to Edmonton, or finding a way, whatever way, from Ellis Island to Franklin Furnace to Glace Bay and then west. I am looking for a way to connect to him, however tangentially.

At the huge flea market near Porte de Clignancourt, we found one booth of tables filled with wooden type, large letters—forty-eight point, or bigger, many of them, suitable for posters—and briefly considered buying some. But we'd need a lot to print anything with (think of how many vowels are used in a short sentence, how often an "s," a "t"...) and the letters would take up so much room in our luggage. I ran my thumb along one letter, an X, and wondered if it was beech. It was hard, and dirty with old ink. Some came off on my hands. For days, the ink lurked in the deep creases of my palms, reminding me of its archaic utility.

In Venice, a death.
O the metaphysics of time: that I could stand at a phone kiosk on the Campo San Pantalon, calling my mother on a Saturday evening in November to reach her as she drank her morning coffee. "I won't lie to you," she told me. "He has a cough that the nurses say means he will probably die this weekend." Her weekend was beginning while mine was half-finished.

When I entered the churches of Venice to look at Tintorettos and Tiepolos, sometimes it seemed appropriate to light a candle for my father. It wasn't my religion but it had been his, and I thought that the light of those tall white tapers was something he would have approved of. A few words only, to ask peace for him, surrounded by images of death and resurrection, and the silence of the long-eyed madonnas. I felt closest to him on Torcello, at the ancient Santa Fosca, its Greek-cross interior echoing the churches of my grandfather's Bukovina. I lit a candle there in the silence, asking for grace.

A day later, he was gone. "Last night," they told me when I called again and reached my daughter, my mother, and my youngest brother at the hospital, packing up my father's clothing. It was morning for them and I was on the Campo again, in darkness, the sombre beautiful buildings lining the canals, while all over the city church bells rang out the hours and gondoliers waited with their elegant black boats. At dinner we toasted my father with red wine and offered each other memories: a camping trip where my father taught my sons to fish, made them pancakes, took them out in his dinghy. The time he found a small wheel and made a child-sized wheelbarrow to complete it. We returned to our hotel where the patron smiled as we open the door, his parrot on his shoulder. "My father died," I told him—to explain the tears. He nodded gravely, made a motion with his hands to indicate that everything turns over, and said he would like to come back as a flower.

Across northern Italy and along the south coast of France, I gazed out the window of the train and thought about legacies. How we are shaped by people we barely know or not at all. Their lives touch our own in the faintest of ways, like the pattern of graffiti on beech bark leaving its mark on damp paper or the whorls of our fingers, texts encoded in our bodies (the dark ink of wooden letters staining my hands). The swarthy pigment of my skin is partly the story of a young man travelling steerage to North America and finding his way to Phoenix where barely a trace of him remains—only a name on a census, and the designations "Single" and "Boarder." And the town itself only a tiny mark on the landscape—a clearing, a monument; though its dead lie in the ground with their cryptic stones. And who knows what messages might be imprinted in the trunks of trees? The names of lovers, a cry of despair or joy, a date with no other note.

Within the next few years, I hope to travel to Bukovina and see its monasteries, those forests of beech trees. I will do this for myself, and for my father, who never indicated much interest in what might be there, if anything. Without the language of the country, I'll try to find something of my grandfather, something as solid as names in a graveyard in Ivankivtsi, or as transitory as the sound of birdsong on an autumn morning. And for my father, I will plant a copper beech beyond my garden, a small pinch of his ashes in the soil if possible, roots nourished by his memory.

Arbutus menziesii
Makeup Secrets of the Byzantine Madonnas

> And take great care, if you want your work to come out very fresh; contrive not to let your brush leave its course with any given flesh color, except to blend one delicately with another, with skilful handling. But if you attend to working and getting your hand in practice, it will be clearer to you than seeing it in writing.
> — Cennino D' Andrea Cennini, *The Craftsman's Handbook*

It was late spring. I don't remember the weather. I was twenty-two years old. We were in his studio, and I was taking off my clothes — jeans, no doubt, and a baggy sweater. A bra. Cotton underpants. I had agreed to recline on the divan in the light-filled room and allow him to paint me.

I wasn't thin. In fact, I was quite plump. The idea that someone found me beautiful made me throw my usual caution about my body to the wind. Or at least my clothing to the floor, before I found a position which I felt I could maintain for an hour or so while J. mixed paints and began the process of putting me on canvas.

The canvas was already prepared, I remember — he favoured a glue sizing made from rabbit skin, and gesso for the ground. He was working *alla prima*, wet paint on wet, hoping to complete the painting in a single sitting. I don't remember if he underpainted, though I imagine there was some basic drawing first, to block out the design. But he wore little glasses when he painted, and he peered at my body through them in a clinical manner, not missing a thing. I thought of my large breasts (heavy even then before years of nursing the children that arrived later in the decade), and the heft of my thighs, and hoped he wasn't disappointed. He wasn't.

It was just that once that I took off my clothes and from that occasion, he made hundreds of sketches, several large paintings that I know of, and even a folio of prints in various media. The second time he met my husband, four years later, he presented him with a set of the prints. I was amused to see myself in poses I knew I'd never affected. I realized how a man's imagination can turn a woman into anything he fancies — wanton, careless as she arranges her hair, gazing over her shoulder, opening her mouth a little as though she is aching for love.

Arbutus has a limited range on our Pacific Northwest coast. It's shade-intolerant and likes well-drained soil, sending its roots deep for water. The trees don't like to be moved. They don't like to be too far from the sea — a tree after my own heart.

Arbutus trees are declining throughout their range, in part because the areas they favour are also areas where people are increasingly drawn. I think in my own community of the recently developed Daniel Point, where huge houses occupy land that was once home to many arbutus (though to be fair, many property owners have included the trees in their landscape plans). I think, too, of Vancouver Island, where retirement communities have taken over slopes once occupied by groves of arbutus and Garry oaks, which are often companions. Like the oaks, they are sensitive to changes in their drainage and root system areas. Fire used to be a factor in creating favourable habitat for these trees, eliminating the conifers that tended to take over, providing too much shade for the arbutus to thrive.

Arbutus are signature trees of a particular landscape — they love bluffs and help to stabilize them with the far-reaching nature of their roots. Think of those cliffs along the coastal highways knitted together by arbutus roots!

Occasionally at a craft sale someone will be selling a small bowl or spoon made of arbutus wood, and I marvel at its clean appearance. The wood is very hard and smooth, with a tight grain, though apparently it dries unevenly and is brittle, making it difficult to use for flooring and bigger projects. I remember sailing with a friend many years ago on Haro Strait, stopping on a small island for lunch (he baked muffins in the galley's wood-burning oven). He was delighted to find an arbutus log on the beach and promptly cut it into short lengths for his stove. He said nothing burned hotter or more reliably.

The Straits Salish on southern Vancouver Island tell a story of a Great Flood, that survivors in what is now Saanich tethered their canoe to an arbutus on top of Mount Newton. I think of that tree, deeply anchored within the bedrock of the mountain itself, hardly concerned by the rising waters all around it, and how such stories speak to the specific, how we are shaped by the familiar made numinous, often by its reliability. Someone had noticed that particular arbutus and remembered it when the waters began to rise. That faith was rewarded. Like the Nuu-chah-nulth story, on the west coast of the Island, which tells of people weathering a flood by tying their canoe to bull kelp (rooted deep in the ocean floor), this speaks to careful attention resulting in survival.

> First take a little dish; put a little lime white into it, a little bit will do, and a little light cinabrese, about equal parts. Temper them quite thin with clear water.
>
> —Cennini, *The Craftsman's Handbook*

I knew nothing about paint. To watch the mixing of colour was alchemy. The palette was a mess of pigments, and J. kept mixing vigorously with a brush. Dab, mix, dab, peer at me through those glasses.

We talked throughout about painting. Our favourite artists—Augustus John, Gwen John (whose work we liked even more than her brother's), Sutherland, Balthus, Picasso, Pisanello. We talked about relationships between artist and model, how for some painters an ikonic image was constant, guided them, served as a deep source of inspiration. J. had always painted dark-haired women, and showed me canvas after canvas, each of the models bearing a resemblance to each other, to me.

I wanted to know the names of the colours—titanium white, ivory black, yellow ochre, ultramarine, cobalt blue, Venetian red, alizarin crimson, viridian green, terre verte, light cadmium red. There was poetry in the colours, their ability to change and alter when nudged with a little of another. And how to organize them on the palette, remembering the order; the dark colours looked similar to me, in dark mass, but a little white dabbed on the edge of the mass revealed blue or dark green.

J. explained the process of taking a good sketch to the canvas, using a grid, using charcoal or a pencil to keep the elements of a design intact. "Don't worry about buttons or eyes at this stage," he said, studying my body, mixing the colour for my nipples.

> Touch up the hair with verdaccio; then with this brush shape up this hair with white. Then take a wash of light ocher; and with a blunt

> bristle brush work back over this hair as if you were doing flesh. Then with the same brush shape up the accents with some dark ocher. Then with a sharper little minever brush and light ocher and lime white shape up the reliefs of the hair.
>
> — Cennini, *The Craftsman's Handbook*

In that particular context, I suppose I was a muse. It took me years to recognize this. A young woman is often the last to recognize her own attractions. I grew up in a family that didn't praise. We weren't unusual. My mother's background was Presbyterian. If I looked in the mirror for too long, or expressed too much interest in my appearance, she would tell me I was vain. And yet — oh paradox! — my mother wanted me to have curly hair like Shirley Temple (a perfect child, with her blond curls, dimples, and a sweet nature). Mine was disappointingly thick, brown, and straight. Never mind, that's why the home permanent was invented. I remember afternoons spent crying in a haze of ammonia as my mother wrapped sections of my hair with paper and rolled them onto small curling rods before drenching my scalp with Tonette. I'd weep as my mother wrapped and my father would say, "That's the price of beauty," and I cried even harder: "I don't want to be beautiful. It's Mum who wants me to be!"

The perms never worked. I never looked like Shirley Temple. I'd have needed short hair to begin with, and someone who knew that one had to begin high on the scalp with the curling rods instead of rolling them from the bottoms of the strands. So I'd have smooth hair from the crown and then uncontrollable frizz for the last four inches.

The perms coincided with Easter, and there was a new dress and hat for the Easter church service. I was mortified at the way the frizz hung below the brim of the straw hat. It took weeks, even months, for the perm to settle down, and it was always a grave disappointment to my mother, and to me — for I wanted her to be happy with how I looked. Yet each year, the resurrection of the Lord prompted the ritual of the perm.

In my early teen years, I'd been very aware I wasn't the kind of girl boys my own age were drawn to. I was a little taller than average — 5'6" — and not thin. I was dark-haired, dark-eyed. One of my best friends was very fair, with blue eyes, delicately pretty. When we went to parties together, she was swarmed by admirers. Beside her I felt gawky and plain. Once I waited in the car while a boy walked her along the waterfront, kissing her in the moonlight while I tried not to watch.

But later, when I was seventeen, I began to attract another kind of attention. There was a man who did deliveries for the pharmacy where I worked on weekends. He'd look at me in a deeply admiring way, seriously, and I felt like he was taking my clothes off in his imagination. He had an old Rolls-Royce, I remember, and often suggested that he drive me home after work. As I lived about three blocks away, I'd refuse him, politely, the way I'd been taught to treat those older than me. He was older than my father, after all, and I was not comfortable with his interest.

One of the pharmacists used to follow me into the lunch room and try to rub against me. He had a crewcut and pop-bottle glasses, terrible breath, and a wife who always wanted a substantial discount when she bought makeup from the cosmetician. I could not imagine kissing him — or for that matter, lying with the delivery man in soft grass somewhere hidden from the road — for surely that was his intention.

But that became the way of it. Older men, as old as my father, or more ancient (my father at that time was about forty-seven), began to show me the attentions boys my own age couldn't, or wouldn't. They weren't all shifty-eyed, in back rooms of a pharmacy, or leering from the magazine aisle while I rang up a customer's order. Some were courtly; they noticed my appearance (hair styles noted and approved of, a dark green dress praised by the father of a young man I dated, my graduation dress of ivory silk commented on by my high-school teachers). They talked to me as though to an intelligent person whose ideas were worthy of being taken seriously.

My father, in contrast, argued with me about everything and looked down his nose at my emerging political beliefs — though I must confess I expressed them aggressively. Coming home from my first opportunity to vote, I said airily, "Well, I guess my vote cancelled yours," having made my mark beside the name of the Communist candidate. Although I knew almost nothing about communism, apart from a brief introduction in my modern world history course at school, I knew that my father hated the idea of anyone other than the Progressive Conservatives as our governing party. He hated Trudeau. Communists, socialists: anyone who mentioned the poor or human rights or the seal hunt was a bleeding heart. He hated anyone who he believed had behaved cowardly or otherwise badly in the Second World War. This included people whose origins were in countries like Japan or Germany or Italy, even if their families had arrived a generation or two earlier. Like ours had.

My father had no interest in art or poetry or any of the things I was discovering could teach me to live as I needed to. When I made plans to travel abroad, he said, "What's wrong with here?" "Here" was my parent's small house, where we crowded around an Arborite table and ate from Melmac plates. My mother made wholesome food but did not go to any effort to serve it nicely, apart from at Christmas; she was too busy with the daily work of a household of six. In my parents' home, listening to music was thought of as "affected," and the notion of a lovely table set with pretty dishes and flowers was "putting on the dog."

The winter I shared a beautiful heritage house with a friend — it was her family's summer home, built in the early part of the twentieth century on its own little cove north of the city — I began to see that houses and meals could be gracious and not simply utilitarian. That house had a shelf of cookbooks that I read like novels; Elizabeth David's *Summer Cooking* is one I particularly remember. From that book, I made a salad Niçoise, delighting in the arrangement of steamed beans and golden-yolked eggs, glistening olives, small creamy potatoes. It was a world away from iceberg lettuce and Thousand Island dressing.

When I thought of my own modest background, I felt a kind of shame in the contrast, as though I was betraying my roots in wanting to surround myself with beauty.

I was living in that charmed house (I dream of it still) when I met a visiting poet from Venice who took me for long, drunken dinners and recited poems in my honour (though I realized later that the poems had been composed for earlier loves; it was convenient to have them memorized when he fed me delicacies, washed down with glasses of red wine) and who proposed marriage during our fourth meal together. Though flattered, I was beginning to realize that he wasn't entirely the man I had been waiting for, and ended the relationship.

Some nights, I stood on the stone terrace while the sea lapped just below me, wondering if I would ever see Venice in my life now that I'd stopped taking the poet's phone calls. He'd call late at night, manic and tearful, almost incoherent with drink, lapsing into Italian as he sang my praises or lamented my indifference. Anyway, it wasn't indifference but fear that made me replace the telephone receiver while he carried on at the other end. I was afraid I had gotten in over my head, and the only way I knew to save myself was to hide.

Only weeks later, he was courting another young woman who did marry him and go with him to Europe, giving birth to several children, living (as I understand it) in poverty while he wrote poetry and translated obscure texts (he was a polymath, among his other gifts).

There were others. The list embarrasses me now, as I see its pattern so clearly. Yet there was something compelling about older men treating me as though I'd been gilded with a kind of divinity. Which I see now was simply youth.

With a very soft, rather blunt, bristle brush take some of this flesh color, squeezing the brush with your fingers; and shape up all the reliefs of this face. Then take the little dish of the intermediate flesh

> color, and proceed to pick out all the half tones of the face, and of the hands and feet, and of the body when you are doing a nude.
> — Cennini, *The Craftsman's Handbook*

This is hard to talk about. When J. painted my body, I felt a strange thrill — the brush stroking paint across my abdomen, and concentrating on my nipples, the dark vee of my pubis. I never made love with him but in this way, he made love to me. I was beautiful under the bristles and sables, flats and rounds. Beautiful against the flowered cloth, the imperfections of my skin eased away by his touch. Outside, forsythia bloomed against the window, so it must have been March. Afterwards, there were glasses of wine and a plate of cheese with a few grapes. It was like talking in bed, after love, although of course at this point we were again fully clothed. Or I was, for he had never taken off as much as a shirt.

I thought about J. (now deceased) and his seductive work on a recent trip to Venice. (How gratifying it would have been to the young woman I was to know she *would* see Venice one day, and not in the company of a man who wept under the influence of red wine and who groped for her body in tears, but rather with her beloved husband.) One of the first places we visited was Santa Maria Gloriosa dei Frari, where I stepped into a sacristy to meet Paolo Veneziano's Madonna. She sits serenely within her altarpiece as she has for almost seven hundred years, holding her son on her lap, wearing a cloak as blue as Giotto's skies, gorgeous with *fleur-de-lis*. Though a doge and his consort are being presented to her at one side of the painting, she gazes away to the saints at her right, not at them, though she is also aware of us looking. Her knees beneath her draperies look very strong. Cennini says:

> There are some masters who...take a little lime white, thinned with water; and very systematically pick out the prominences and reliefs of the countenance; then they put a little pink on the lips, and some

"little apples" on the cheeks. Next they go over it with a little wash of thin flesh color; and it is all painted, except for touching in the reliefs afterward with a little white. It is a good system.[1]

I don't know if this was Paolo Veneziano's particular method, but he did something beautiful with his Madonna's cheeks. And her eyes are ravishing.

Then take a little black in another little dish, and with the same brush mark out the outline of the eyes over the pupils of the eyes.[2]

I had a single eye pencil in my toiletries bag at the Albergo Casa Peron on Salizzada San Pantalon and try as I might, I couldn't outline my own eyes in that intriguing way.

I saw those women everywhere, their eyes out of Cennini. Paolo Veneziano's *Coronation of the Virgin* at the Galleria dell'Accademia, his *Madonna and Child Enthroned* and another gem by Giambono, also a coronation. Bellini's later *Madonna of the Little Trees*, her downcast eyes not made up as dramatically as the Byzantine fashion, but still wonderful, more human. By the early Renaissance, the Madonnas were actual women rather than the primarily stylized images of devotion modelled on the glorious golden ikons coming to Italy from Constantinople and Greece.

In that fall of 2009, Venice was still full of those women, dark-haired and dark-eyed, their skin glowing. I'd see a woman standing on a bridge over one of the canals, talking into a cellphone, her shoulders wrapped in a shawl of rich carmine or China red, balancing on one high heel, turning the toe of the other sleek boot this way, then that, her free hand brushing her hair back from her face. And I, full of Titians and Tintorettos from a day of entering churches and hardly believing that the next might contain more beauties than the last, would recognize her profile, her animated hands.

We went by train to Padua for a day and walked from the station to the Scrovegni Chapel. The process for admission was complicated. We bought our tickets at the civic museum and walked over to the chapel; it was brick, very plain and unadorned. We hadn't understood that small groups, no more than twenty at a time, were assigned a specific time for admission, based on reservations; but because the group waiting to go in had fewer than twenty, we were allowed to wait with them. We had to sit in an antechamber for half an hour while dehumidifiers hummed and did their work, quietly absorbing the fall moisture from our clothing. We were shown a film on the history of the chapel. Then we were ushered into the chapel through a series of doors designed to keep damp air at bay. We had twenty minutes to look at the astonishing painting cycle begun (arguably) in 1303 and completed by the consecration date of March 25, 1305.

It was like entering a novel, episodic in form, providing scenes from the lives of Joachim and Anne, then their daughter, Mary, her husband, Joseph, and her son, Jesus. It was necessary to find our way through the narrative quickly, the clock ticking those twenty minutes away. Like entering a novel: beginnings, plot development, rising action, a denouement, a death, then resurrection. The panels were so richly animated. Everything was in them — serenity (Anne in her bedroom receiving news of her impending maternity from an angel climbing in through a high window, its wings barely clearing the lintel, while outside her handmaiden spins yarn) in contrast to an earlier panel of Anne's husband Joachim meeting shepherds after being driven from the Temple of Jerusalem for his unworthiness (i.e., he hasn't been blessed with fatherhood). In his panel, a small dog leaps at his feet, and the sheep mill about while the shepherds look puzzled at his arrival. Later, we read the flight into Egypt, Mary and her infant on a remarkably placid donkey, her cloak having lost most of its azurite pigment from the centuries of damp.

("If you wish to make a mantle for Our Lady with azurite," advises Cennini,

> …begin by laying in the mantle or drapery in fresco with sinoper and black, the two parts sinoper, and the third black. But first scratch in the plan of the folds with some little pointed iron, or with a needle.

> Then, in secco, take some azurite, well washed either with lye or
> with clear water, and worked over a little bit on the grinding slab.
> Then, if the blue is good and deep in color, put into it a little size,
> tempered neither too strong nor too weak...Likewise put an egg yolk
> into the blue; and if the blue is pale, the yolk should come from one
> of these country eggs, for they are quite red. Mix it up well.)[3]

Mary's cloak may be flaking from the damp but her eyes are beautifully long, and outlined in black. Her husband looks back in awe and concern. And perhaps the most dramatic scene of all, in the garden of Gethsemane: even from our distant place on the floor, centuries later, I could see the furrows of Judas's brow as he kissed Jesus in that moment of betrayal, Peter slicing an ear from a soldier, raised lances, a horn blasting, torches flaring in the night. And everywhere, those dark Byzantine eyes.

J. made me a little book, hard-bound in black leather. I was going away, in part running scared from his attentions, in part to see if I might make a life for myself, alone, in order to test my writing gifts—I thought they needed to develop in isolation, or that I needed to develop in isolation, and for a time it worked. He hoped I might try to paint, because he was convinced I could do anything. (The little scribbles I produced in his company, using his charcoal and fine papers, were praised far beyond their worth.)

The book set out to teach me basic painterly skills, beginning with materials, preparing the support, taking an image from sketch to painting, demonstrating with arrows and crosshatching how to keep a composition dynamic, mixing colours on a handy plate or piece of glass. His first intention (I believe this) was to offer me quick lessons in the skills I'd need to paint. But it quickly filled with drawings of me—my face, my body (in poses I'd never have agreed to), my clothing peeled away like bark. I turn its pages now, a woman in her fifties, and understand something about obsession. Maybe even love.

Out the window of my study, an arbutus tree stands in a bed of periwinkle. Warblers dart among its blossoms in spring, eager for the tiny insects that are drawn to their honeyed scent; band-tailed pigeons visit in late summer for its berries. It's like a patient beautiful woman, arms open to the sky. And looking at the tree, I'm reminded of how age too can be peeled away like bark to show the smooth new layer, waiting to receive its first experience of sun and rain, the light feet of birds, a young snake curled at the base, listening with its tongue.

In Venice, we looked at hundreds of paintings over two weeks. I noticed colour like never before. Was it November's grey skies and the dark water of the canals that made things so vivid by contrast? The ancient buildings with their peeling paint more beautiful for the wear? No matter. I would enter a church or a *scuola* and the reds would ravish my eyes. The dense warmth of the yellows heated the chilly interiors like embers. And skin glowed like the flesh of the young, as though lit from an inner source.

 I loved the carmines (from cochineal insects), the Venetian red (derived from red iron oxides), ultramarines, azurites, Egyptian blues, purples from indigo and madder, malachite and verdigris (from acetate of copper) greens, lemony orpiment and Naples yellows, burnt Siennas and umbers, the heavy white of lead and the lighter chalks and gypsums, and the carbon blacks (from burned bones, soot from lamps, charred remains of vines). There was a shop we passed often in the Dorsoduro where powdered pigments filled a tray in the window, the colours crying out to be purchased and mixed.

 What would they have been mixed with? Oil, of course: linseed, walnut, or poppy. With solvents to dilute them for translucent glazing. Different colours dry at different rates, so that would be considered when mixing — turpentine, or a little more oil, perhaps, added to the Alizarin crimson and some of the yellows which dry very slowly. These ingredients would be added much more sparingly to the ochres and Venetian crimson (depending on the schema of the work itself) because they dried faster. An artist in the heat of creation, wanting layer after layer of transparent colour to intensify hue and create the optical effect of warmth and depth, would want the layers of paint to dry quickly enough to allow the momentum of the work to be sustained (unless there was something to be gained by working *alla prima*).

In the past few years, inventories have emerged from the dusty depths of the state archives in Venice which reveal that sixteenth-century painters had convenient access, through sellers of artists' pigments, to the raw materials used by glassmakers and dyers — finely ground particles of coloured glass.[4] The use of blue smalt, or finely ground blue cobalt glass, was already known about in some fresco work; but further analysis, using scanning electron microscopy and energy-dispersive spectrometry, of the work of Lorenzo Lotto and Tintoretto revealed presence of other colours of glass, the silicas and irons allowing for an expansion of colour choice. The glass was used with pigment in very thin glazes that resulted in transparency, vibrancy of colour, and luminosity, qualities associated with the work of these artists. The glass also acted as a drying agent, fundamental in facilitating the application of many layers.

Those grey days in Venice, I made my way along quiet back streets among carnival masks and shops resplendent with pastries as beautiful as sculpture to Madonna dell'Orto, Tintoretto's own church in the Cannaregio where his canvasses filled the space. *The Presentation of Mary at the Temple*, the *Sacrifice of the Golden Calf* — such colour and drama! To Sant'Alvise where the Tiepolos brought tears to my eyes — Christ laden with his cross on the road to Calvary, his flagellation, the crown of thorns. And Tiepolo again as we sat in the Scuola Grande dei Carmini to listen to *La Traviata* on a dark night, light casting its spell on the flesh of the heavy-eyed Mary on panels, the plump Violetta singing on the stage.

Near Arbutus Point
The list embarrasses me, but I remember one older man, an amiable satyr, who took me to a hidden beach in moonlight and laid me down on his coat, spread over cool sand. What he did with his tongue was miraculous. No boy I'd dated ever suggested such pleasures were possible. They brought out their wallets, the leather imprinted with the ring of a condom, those badges of honour—and saw no reason why anyone needed to go farther than the back seat of a car, parked at Beaver Lake or the end of my parents' road. Not to remote beaches in moonlight where arbutus trees rustled in wind and a bald man made me cry out in surprise.

> But I will tell you that if you wish to keep your complexion for a long time, you must make a practice of washing in water—spring or well or river: warning you that if you adopt any artificial preparation your countenance soon becomes withered, and your teeth black; and in the end ladies grow old before the course of time...And this will have to be enough discussion of the matter.
> —Cennini, *The Craftsman's Handbook*

More than thirty years ago, I removed my clothes for an artist, each layer—the baggy sweater, jeans, cotton underpants, lace bra—flung to the ground in careless abandon of a self I hoped I could transcend, on canvas if not in fact. What he wanted from me wasn't physical exactly. It was what men often hope to find in a woman's presence that makes itself known in her body. Before that, and afterwards, there were others who found this in me though I was puzzled by their conviction that I had something they needed. Occasionally I recognized it in poems written for other women, for instance the beautiful "On Raglan Road" by Patrick Kavanagh.[5]

I gave her gifts of the mind I gave her the secret sign that's known
To the artists who have known the true gods of sound and stone
And word and tint. I did not stint for I gave her poems to say.
With her own name there and her own dark hair like clouds over
fields of May...

— and also in poems written for me, most of them by the poet who became my husband. Time provides such clarity and from this great distance I wish I'd been more easy with the role in which I'd been cast. It troubled me then because I thought I was at fault, that I wasn't worthy of the kind of relationships my friends were entering into. Now I can honestly say it was a privilege to (however briefly) occupy the imagination of a man who caressed my skin with brushes of hogs hair and sable, and who filled a small book with my image.

I am memorized on canvas, on paper, a Madonna without the beautiful long eyes or that wise serenity. And I look down from the wall of our living room, a poster girl in a robe of Alizarin crimson. A poster girl whose face was underpainted in terre verte (an unctuous earth pigment, taking its green from hydrated oxide of iron), and then left. Surely this was an exercise in *verdaccio* and my painter intended to use the grey-green to establish the values for painting my flesh with successive layers of glaze? Unfinished, or abandoned, or given up to dull green earth. It took time, but I came to love this version of my younger self, uncharacteristically elegant, her skin echoing the new bark of the arbutus framed by the large window. And there's another portrait, a dreamy girl with flowers in her hair; she is wearing a blue wool vest I sewed from fabric bought at Capitol Iron in Victoria. She hangs high in the stairwell and gazes down as I descend the stairs each morning, dishevelled and eager for my coffee. I'm not sure if she sees me or sees through me. And if the obverse is also true.

There are arbutus trees at Francis Point, a grove of them, leaning out to sea, wanting to partake of the cool air off the water on summer days. Mount one and stretch your body along its length. Has there ever been a tree more seductive to the touch? Has there ever been a trunk, peeled of its bark and new, more like the smooth torso of a beloved? Without mark or blemish, asking us to run our hands along its taut muscularity? The underwood is chartreuse, radiating light.

How many times do we shed our outer layers in a life? How many times expose our tender new skin to the world, soft as the soles of a child who has never touched the earth? Looking out my window, I see the bark curling from the arbutus on the south side of my house. Like paint peeling from an old surface, we hardly notice it but are drawn to what's revealed underneath. Steaming the bark with the pale bulbs of camas would turn them pink as young flesh, beauty for the eye and the palate.

Postscript
Even the red-breasted nuthatches that visit my feeders have a black line elongating their eyes to make them as elegant as Veneziano's Mary...

"May I help you?" asked the nice woman at the cosmetics counter in Shopper's Drug Mart in Sechelt. "I'm not sure," was my reply. I explained about the eyes of the Byzantine Madonnas and without even raising her own well-shaped eyebrow she sat me on a stool, brought out a pot of deep brown powdered eye shadow and a thin brush, and tilted my face up with one hand on my chin. Deftly she brushed a thin line along my eyelids, top and bottom, and gave me a hand mirror to look in. Well! I certainly don't have the long almond eyes of those ikonic women, nor do I share their beautiful serenity; but I was pleased to see a quality of which Cennini might have approved. *Then take a little black in another little dish, and with the same brush mark out the outline of the eyes...* I bought a pot of the shadow and a brush and have been trying ever since, without success, to replicate the effect. Which goes to show that some have the sure hand of an artist, and some don't.

Populus tremuloides
Cariboo Wedding

Leaves: Alternate, deciduous, simple, broadly egg-shaped, kidney-shaped or circular, 2.5-9 cm long, 2.5-8 cm wide, the bases rounded to slightly heart-shaped, smooth, finely toothed and fringed with long white hairs, upper side green, lower side paler; leaf stalks 2-7.5 cm long; buds smooth[1]

It was a long way to drive for a wedding, from our home on the Sechelt Peninsula to the Nazko Valley, west of Quesnel. We left on a Thursday, in late July. Up early, second ferry from Langdale to Horseshoe Bay, Sea to Sky Highway to Pemberton, then east on the Duffey Lake Road as far as Lillooet the first night.

In Lillooet, we could see flames on the mountain behind the town. Helicopters were swinging buckets of water over the flames, clouds of smoke billowing into what had been a clear sky as water hit fire. The woman at the post office said part of the town was on evacuation notice but when we said we'd been thinking of staying a night in Lillooet, she quickly said, "Oh, you'll be fine. It's only the outskirts that need to worry." A small town in depressed times: every tourist dollar counted.

We wandered around and looked at things. The main street of Lillooet was Mile Zero of the wagon road leading miners from Port Douglas to the goldfields during the 1860s (many place names on the route evolved from the stopping houses along the way: 70 Mile House, 100 Mile House, the 108 Mile Ranch, etc.). The beautiful Miyazaki House was elegant in its shady garden, trees hung with bright apricots. Farther down Main Street, the museum featured in its basement a dusty approximation of Ma Murray's *Bridge River-Lillooet News* print shop (Margaret "Ma" Murray and her husband George launched the newspaper in 1934). We couldn't decide at first whether to stay or move on to a town without a fire at its back. Clinton, maybe — or Cache Creek. But we liked Lillooet. I felt that there was a story in the plantings around the Miyazaki House — that the fruit trees, the big cottonwood, and the lilacs might figure into this memoir I was writing about trees. So we reasoned that the fire couldn't reach the town overnight, and it was likely we could sleep peacefully in a room at the Mile 0 Motel without being roused by loudspeakers to pack our bags and leave within twenty minutes.

Our room at the Mile 0 was adequate, though spartan. A few thin towels in the bathroom and two Styrofoam cups by the coffee maker. Still, there was a view of the Fraser River if we stood on the small balcony with sliding doors to let in air. If we stood outside the front door of the unit, we could watch the drama of the helicopters, one after another, rising from the Fraser River with their buckets.

We loved the Miyazaki House. It was built in the late 1800s (a heritage report issued by the District of Lillooet indicates the construction was between 1878 and 1890) in the Second Empire style, with a mansard roof. Built for Caspar Phair, a merchant and gold commissioner, the house, with its gracious lawns and gardens, was a centre for social activity well into the next century. During the Second World War, it became the temporary home and surgery for Dr. Masajiro Miyazaki, a Japanese-born Vancouver osteopath who'd been interned at nearby Bridge River when Canadians of Japanese ancestry were evacuated from the Coast. Dr. Miyazaki's medical skills were needed in the area and he was recruited as Lillooet's coroner when the town's doctor died during the war. After the war, Dr. Miyazaki was able to buy the house from the son of Caspar Phair and lived in it, raising his family there, until 1983, when he donated the house to the town of Lillooet.

Walking the main street of town in the evening, after a good meal at the Greek restaurant, we saw people on every block or corner, all of them looking towards the mountain. It was a Thursday, not a weekend, but it felt almost festive, walking the sidewalks of Lillooet where residents had set up lawn chairs with small coolers full of beer and soft drinks for the children. Some people kept binoculars focussed on the fire. We bought ice cream cones and ate them on the patio of a little café; around us, customers sipped cold drinks and talked of the fire. I expected to smell smoke but it was just a faint whiff, a rumour. Several times in the night, I went outside to see a deep orange glow against the dark mountain. Other motel guests were watching too, an uneasy fellowship. I imagined I could hear the crackling of flame, fanned by wind off the Fraser River, but in fact it was strangely quiet.

We were heading to a wedding so we drove off before 7:00 a.m. the next morning; the helicopters resumed the water-drops at 6:00 a.m. so we didn't need a wake-up call. Our car was covered with a fine film of ash. By now we knew that the mountain was Mount McLean, and that there were no roads to where the fire was raging on its slopes which meant the blaze would be fought mostly from the sky. As we drove over the bridge across the Fraser River, we saw the smokejumpers' campsite on its banks, a village of tents that had sprung up overnight. Young men were crawling out of tents, pulling on red shirts with the fire-service badge. I thought of another MacLean — Norman — and his wondrous book about another fire fought by young men, the 1949 Mann Gulch fire in Montana, a book of Shakespearean power, which ended in tragedy. In my heart, I wished these guys courage and luck.

The Pavilion road was beautiful in the cool of morning. In the valley bottom, there were trembling aspens, their heart-shaped leaves fluttering while horses grazed above the river in distant fields. Hawks watched from telephone poles. And everywhere a soft wind, the smell of dry earth. It felt like morning at the dawn of the world, the old gods walking those shimmering fields, attended by rustling leaves.

I had my copy of *Plants of Southern Interior British Columbia* in the car and looked up trembling aspens. The delicate movement of their leaves, dark green with pale undersides, made a sibilant music. The white bark was a perfect contrast to the cinnamon trunks of the ponderosa pines. What would the book tell me about their ecology?

"Reproduces mainly from root suckers following disturbances, such as cutting or fire."[2] Well, this was a landscape vulnerable to fire, I thought, looking behind us to see the smoke from the McLean Mountain inferno while the radio news reported that its size had doubled overnight. I remembered my childhood among the fire-shaped Garry oak meadows of southern Vancouver Island and saw some similarities to these dry expanses of grass. There is also a history of controlled burning by the Stl'átl'imx people[3] to enhance plant resources, notably berries and roots.

While we drove, we were listening to Lorraine Hunt Lieberson sing Handel and Bach arias.[4] Two years into my singing lessons, I was happily accompanying her while John drove; our windows were open and warm air carried the tang of sage into the car. The aspen leaves were rustling and when we passed a grove of them by the side of the road, we could hear their whisper. It was an idyllic drive, Lorraine (and me) singing,

> There in myrtle shades reclined,
> By streams, that through Elysium wind,
> In sweetest union we shall prove
> Eternity of bliss and love.

Such a potent landscape — the rock formations and pines and golden stretches of grass. We passed small creeks entering the Fraser River and shelves of rock which were fishing sites for the First Nations people of the area. These sites were owned by extended family groups and had names passed down through the generations; sometimes a place speaks its own name even now: the site known as "shady rock" or the site known as "foaming." The sockeye were running, heading for their spawning beds near Horsefly and Stuart lakes, and I expected to see people out with nets but didn't. We did see a pair of coyotes on a small hill, taking the sun, and hawks swooped as chipmunks raced across the highway.

Our destination was Quesnel. We'd reserved a suite at the Talisman Motel, and we didn't know what to expect. The marriage of my niece Lisa and her beau Chad would take place on Saturday afternoon at Rainbow Lake in the Nazko Valley, an hour and a half from Quesnel itself; my oldest brother and his wife had their home on the lake and their children — Lisa and her three brothers — had been raised in the valley. Then everyone would return to Quesnel to the Seniors Centre for the reception that evening. It seemed prudent to stay in the town, walking distance from the reception.

Although it was hot when we left home, it was even hotter in the Cariboo, a different heat from the bracing dry warmth of the Lillooet area where we'd spent the previous night. Driving up the long highway, stopped at various intervals for road work (the smell of new asphalt sickening in the heat), we were glad to have air-conditioning in our car, along with bottles of water and bags of fresh cherries to eat at the construction stops.

My family drove this highway when I was a child, on our way to Edmonton to see our grandmother. (There was a faster route but we took this one if my father wanted to pick something up in Prince George — a gun part, maybe — or if my parents wanted to visit old friends in Clinton or Williams Lake.) They were long trips, my father wanting to make the most of daylight, so we'd break up our camps shortly after dawn. He'd pack the station wagon while our mother made sandwiches for later in the day. My brothers and I were delegated to walk the dog and make sure she peed before she was loaded into her spot in the car.

I recalled camping trips in this country with our children when they were small, our own dog panting in the back of our van as we drove the long highway to Bella Coola or Prince Rupert. I'd forgotten the spruce forests along the highway and the beautiful grasslands near Williams Lake, the aspens in small groves, often with horses grazing among them; yet they felt deeply familiar as we drove north to the wedding.

We passed the Sandman Inn on our way into Quesnel, where my younger brother Gordon had first suggested we stay. We were meeting him, and the idea of a shared suite seemed sensible; we could have a visit (we so seldom see him) and we could drive out to Rainbow Lake together the next day. My brother had stayed there for another family wedding a few years earlier and praised it for the adjacent bar and grill, but we were glad when we saw it that we weren't staying there. It was located south of town on the side of the highway. "Can we walk nearby?" we'd asked him, and he confessed it was not really that kind of place. We knew from the description on its Web site that the Talisman Inn was above the Fraser River and near a walking trail that circumnavigated Quesnel. I imagined the three of us walking early in the morning and updating one another on our lives. The Sandman Inn, next to the big parking lot for Extra Foods and Walmart, looked like it could have been anywhere in North America.

And when we located the Talisman, we felt lucky. It felt like we were somewhere in the middle of a western town. It was an older motel but very well-kept, with huge baskets of flowers, a block from the river where that trail meandered along its banks for some 5.5 kilometres. When we checked in, the woman at the desk told us that a complimentary breakfast was served in the lobby each morning. Inside, the rooms were fresh and clean, the air conditioning units were quiet, the bathroom stocked with huge fluffy towels, the kitchenette large enough to prepare meals if one wished. We didn't, but we'd brought food to have with a glass of wine once my brother arrived from Vancouver (he was doing the drive in one long day), so we stocked the fridge with olives, cheese, hummus, a bottle of excellent Pinot Grigio, some beer, and cider.

> Trembling aspens can occur in huge, long-lived clones that may be thousands of years old.
>
> —Parish, Coupé, and Lloyd, *Plants of Southern Interior British Columbia*

It was years since we'd been in Quesnel. When our children were young, we passed through a couple of times on family camping trips. I had no recollection of the long bright plantings of petunias everywhere, and the leafy parks. We'd always been in a hurry, it seemed, and kept our eyes open for Highway 26 to Barkerville, or the road west to my brother's ranch in Nazko, where we visited his family two or three times in those years. Quesnel itself is a small city with a rough reputation though it has seen significant history; it was the commercial centre during the Cariboo gold rush, a role now commemorated with a huge gold pan welcoming visitors to the city. Lumber mills and ranching formed its current economy and a wild summer festival, Billy Barker Days, brought in tourists.

It was hot, in the mid thirties, but we walked into the centre of town to explore a little and see where we might have dinner that evening with Gordon. We saw an Italian place and a Greek restaurant where we knew we'd get the predictable but always tasty plates of souvlaki with roast potatoes and a heap of rice, a mound of salad dense with raw onion, topped with feta. There was a Mr. Mike's—I hadn't seen one of those in years though they were standbys of childhood, one of the only places my family would go to on a Friday night if my father felt he could afford to take a family of six out for a meal. Several places proclaimed Western and Chinese menus. The smell of charred burgers wafted out of the Dairy Queen. We'd eaten Greek food the night before, so were prepared to lean heavily on Gordon to try the Italian restaurant.

How strange, I thought, that my brother from Vancouver and I, from the Sunshine Coast had to arrange to share a motel unit in Quesnel in order to spend some time together. When John and I would pass through Vancouver, we'd think of Gordon and his family; but we were always on our way to distant pastures. Sometimes our destination was the airport, to greet children or fly away ourselves. Or else we'd come to the city with theatre tickets and plans for an intimate dinner together before or after the play. Gordon's family's roads led south, to Seattle, where a daughter and beloved grandchildren live. Or Brazil, to more family. Never mind. It was great to hear his key in the door and wonderful to sit in the cool room, drinking wine and toasting the next day's happy pair.

"Why would anyone, particularly in a town like Quesnel, name their restaurant *Penisola*?" I wondered, while tucking into a plate of fairly good pasta. I was imagining the graffiti already, the jokes in the nearby pub. Gordon and John were thoughtful over their indifferently breaded and fried veal, and then both of them guffawed. There was no way to make it sound nice. The red wine was robust and we drank a lot of it, returning to our motel to watch the football game (them) or read in bed (me). We made our plans for the morning: whoever was up first would make coffee and go to the lobby for muffins and bagels. Then a walk along the river, the trail meandering through parkland and behind industrial areas, by aspens and birches, tall firs and spreading Saskatoons, talking about everything under the sun, before showering and dressing to drive out to Nazko for the wedding at 2:00 p.m.

The first time I'd driven to visit my oldest brother Dan, then living in the teacherage at Nazko as a first-year teacher, it took me more than two hours to manoeuvre my small Datsun over the rough road. In certain areas, it was as runnelled and boggy as a corduroy road through a marsh. I had a friend with me, some produce from her family's farm in the Fraser Valley, and a big box of apples purchased along the way at a roadside stand in Spences Bridge. My friend, a classmate from the University of Victoria, was as curious as I was to see the country west of Quesnel; we studied the map I brought with me over coffee north of Hope and marvelled at the thin blue scribbles of rivers all over the page.

As it turned out, we loved it. It was early October and I remember the aspen leaves were turning and trembled, golden on their stems. It was on that trip that someone told me the Native people called it "woman's tongue" for its soft incessant noise, a notion I recognized later when reading the Scots poet Patrick Hannay (a contemporary of Evelyn):

> The quaking aspen...
> Resembling still
> The trembling ill
> Of tongues of womankind...[5]

The air was so crisp, a perfect companion to the apples we ate as we drove, windows open a fraction of an inch for its sweetness.

    The road has vastly improved since then, and the drive out to the wedding was smooth. We saw a moose in a swampy area and many birds — hawks, in particular. The turnoff to my brother's home on Rainbow Lake had balloons on a post, and a sign — Lisa and Chad's Wedding — to lead us in. The garden was lush (my sister-in-law Linda confessed to weeks of hard work to get it ready for a wedding), there were chairs set up near a log shed with an arbour erected to frame the bride and her groom, children ran in the grass and dogs ran with them, glad of the company. Guests wore clothing appropriate to their notion of a wedding. Gordon changed into the suit he'd brought in the trunk of his car, I had a long simple dress of linen and silk, John wore his best shorts and a summer shirt. Lisa's mother Linda changed quickly from her cut-offs and T-shirt to a dress of oyster satin and looked as young as a bride herself. Dan wore dark jeans, a western shirt, and suspenders to hold up his pants over a prosperous belly. The father of the groom was identically clothed. Those who'd driven the long road from the lower mainland or Vancouver Island were more formally dressed than those who lived in the Cariboo, where a clean shirt or a cotton sundress served as festive wear and was just right in that weather. John was the exception, having refused to even consider long pants and a sports jacket in the heat of July.

When everyone was seated and the music began, the wedding party began to promenade towards the arbour. The wives of Lisa's brothers were resplendent in satin; small boys in suits pulled even smaller boys in wagons, their shirts untucked and clip-on ties askew; Lisa's daughter, strewing flowers as she walked to the arbour, was pretty in pink; and then the bride herself, strong and beautiful on her father's arm. The minister, a small man in a leather vest with a big crucifix around his neck, spoke overly long about sin and God, but then Lisa and Chad were kissing and the cameras caught every moment from every possible angle. Guests stood on the long verandah, out of the heat, drinking cool water and eating sandwiches and fruit.

Calls went out for family photos. Gord and I looked at each other expectantly, straightened our clothes, and began to rise, to walk out to join the laughing and jostling extended family gathered by the flowery arbour. We quickly realized that the Kishkans being summoned weren't us so we casually walked down to the lakeshore instead, pretending it didn't matter. There were calls for Kishkan girls — on this day, in this place, these were the bride, the wives of her brothers, and her mother.

    The distance between our home on this coast and this home in the Cariboo was identical; and the visits far too occasional on both sides; but there is nothing like a wedding or a funeral to both exaggerate and — possibly — bridge that distance. We hadn't kept in touch. It was as simple and as complicated as that. While the cameras flashed and arrangements of Kishkans and Sandfords continued, we headed back to Quesnel to shower and rest before the reception that evening. We were quieter driving back than we had been earlier, but it was hot and the day had already been long and full. Trembling aspens in graceful groves beside the road reminded me that there is more to families than what we see on the surface. Like those trees, my brothers and I shared both rootstock and memory, tangled lines the casual observer would not see nor understand. The absences in the family photographs would be invisible to almost everyone.

How relieved we were to see the chilled wine bottles being placed on each table when we entered the Seniors Centre in Quesnel for the reception at 6:00. It was still hot, inside and out. My nephews, Lisa's brothers, moved affably from table to table with corkscrews, opening the bottles of peach and melon wine, each with its celebratory label: a photograph of the bride and her groom. There were also glasses of dry white or red wine or bottles of beer available from the cash bar and we sat at our table on the edge of the room, the three of us on the outside of a table for eight, so we could see into the crowd, saving room for any others who might arrive late and need seats (no one did). The music was country favourites, perfect for the location and event. The bride and groom swept into the room to strains of "Jackson," while the guests rose to their feet and cheered. This was not an uptight country club event but a robust celebration of a beloved daughter / granddaughter / sister / niece and her childhood friend, now husband. Both had been through earlier relationships — Lisa's produced two beautiful children — and then rediscovered each other in their thirties.

The food was plentiful and good. Roast beef, chicken, perogi rich with butter and onions, and salads; and the speeches were full of love and every possible wedding trope: fairy tales, knights in shining armour, happy endings. The families had known each other for the entire lives of their children, so good-natured insults were traded loudly across the room by the two fathers-in-law like something out of a hillbilly movie, each man straining at his suspenders. Except it felt genuine. My nephews spoke of their sister with humour and an astonishing tenderness.

We stayed long enough to watch the bride dance with her father, then her new husband. The atmosphere was happy and loud. The beer was popular and no one bothered with glasses. The mother of the bride in oyster satin lifted a few herself, and the children raced around on the dance floor in their festive clothes, shoes abandoned. I wished my own children had been present, but they lived too far away—and in any case, hadn't seen their cousins in years. We'd visited a few times when they were small, camping at my brother's place, and taking canoes out onto Rainbow Lake. But as they grew, it was in another direction, needing a different kind of nourishment, though of course all of us shared that original root-stock, its dense spreading tangle.

Walking back, I was surprised at my tears. Surprised at my sense of otherness at a family occasion, where in all honesty, I hadn't even expected to feel like I belonged. To see the Kishkan girls in their formal dresses, flowers in their hair, and not to hear anyone say, But wait, there's another girl, let's not forget her. Though I am now in my fifties and hardly girlish.

Surprised, having driven this long distance, to find out how far I was away from my own true family, the brothers I had grown up with, all of us polite and glad to see the others but almost unknown to one another in our later years.

There had been ghosts at our table at the reception, shadows in the empty chairs. A girl and her brothers, their eventual children. Their parents.

At the Talisman, I thought about families and their journey from original intimacy. I bathed with my brothers, wore their hand-me-downs, learned how to throw a baseball at their instruction. I travelled across Canada with them in a station wagon, sleeping at night in our blue tent; we roasted marshmallows on sticks in glowing coals, pulled bottles of cream soda from icy water at gas stations from Field to Drummondville to Dartmouth. My father photographed us, lined up by age, in front of historical cairns and restored farm implements, all across the country.

I knew my brothers' scent in the dark of the tent—weedy hair from swimming in lakes, smoke from the campfire. When we reached a new place, during the years of my father's transfers, we were company for one another, shared our comic books, played hide-and-seek in teams—older two versus the younger two. Once, when an older kid threatened me in Halifax, my brothers sought him out and let him know that they would defend me, no matter what.

We had camped on St. Mary Lake on Salt Spring Island and taken our father's rowboat out in the mornings to find abandoned homesteads, imagining our lives into those houses. Later, our bathing suits hung on the line my mother strung across the campsite, damp intimate reminders. Yet I have barely seen them for twenty or more years—a day here or there; once for a funeral; twice for anniversaries.

And there was one brother missing from this scenario, who has taken himself far from the rest of us, who has severed his branch from our particular family tree. Yet a tree remembers its missing limb, grows protective layers of bark over the scar. I remembered what I'd read about the trembling aspen, that it "reproduces mainly from root suckers following disturbances" and hoped that this might also be true of our story.

There was a breakfast next morning, at the riverside home of Chad's parents. A huge platter of pancakes, kept warm with aluminum foil, sausages, scrambled eggs, a large jar of saskatoon berry jam, syrup, an urn of coffee. The big tray of Nanaimo bars and brownies that someone forgot to put out at the reception the evening before was emptied in minutes. Dozens of people gathered. The lawn sloped to the Fraser and it was fine to stand there and watch the water heading down to the coast. This was an opportunity for family members like us who'd travelled a distance to visit with each other, catch up. Although our absence was noted at the previous weddings of my nephews, I think our presence this weekend was hardly noticed. People were polite but not curious.

A little moment reminded me of how far my own children were from Dan's children: someone from his wife's family wondered when the next family wedding would be. Much speculation about how long it would be before everyone gathered together again to celebrate a marriage. In my presence, my brother suggested each niece or nephew or cousin once or twice removed and how likely it might be that they would be next. From the fringe of the conversation,

I smiled — but my heart kept asking, Why wouldn't my children count as family? Silly question. Of course I know why. They are barely known.

We said our goodbyes, all of us promising to get together sooner rather than later, and began our drive south, to Clinton where we spent the night. It was so hot. We settled into the Nomad Motel, then walked to the local pub for cold wine and a burger. Later we visited the Clinton Museum, and I looked at the displays of ancient farm tools, mining equipment, and photographs of pioneers long since dead but also vitally present in the implements they'd cared for and left.

Somehow my muddled feelings of remorse for letting my brothers drift from me as they had and pleasure at having spent time in their company and sadness for not knowing them better eased into something like reflection at the passing of time and water and hills of pines, some dying of beetle damage and others scorched by fire, but small groves surviving. Nothing so dramatic as fire had caused our disturbance and among the trembling aspens, the leaves whispered of possibilities.

What is nostalgia but a longing for a time and a place, a hope to return there? That place was not my brother's log house on the shore of Rainbow Lake, though some elements of our relationship were evident there: a canoe tied to a dock, swimsuits on the rail, fishing poles leaning on a shed. We were as necessary and as obsolete as ploughs and horse collars, a pick and shovel from a forgotten claim where someone had intended to dig deep for riches and was disappointed, though perhaps the ore was just a little farther down.

There was a story in the Miyazaki House which I knew I'd pursue. Stories in the Clinton Museum I'd think about and reflect upon. But the story of families drifting apart and then momentarily brought together for a wedding or a funeral — where is the place for that? It's a story as old as time, as old as the memory of travelling through the canyon in a station wagon watching the Thompson River on its inevitable progression towards the Fraser, then the sea.

Sitting on my hands in excitement at the prospect of camping or ice cream at one of the fruit stands along the highway, I'd listen to my brothers bicker and joke while our father shouted impatiently for quiet; the snap of the cigarette lighter pulled from its socket. Back then, I was the only Kishkan girl, hair in pigtails, passed-down jeans rolled up to mimic pedal pushers. Over the course of a summer trip, my brothers sang endless versions of "The Quartermaster's Store" in voices that rang, then cracked and descended an octave. I never dreamed then that a wedding might be the only thing to bring us briefly together again...because it was impossible to imagine a world without them.

Arboretum
A Coda

I am thinking about how the world changes as we sit by our windows, chop wood for our stove, take our familiar walks up the mountain or through the winter woods. Driving out for groceries or meals with friends, we notice the new houses, a recent road leading down to Oyster Bay or carved into the side of Mount Daniel; and on summer days we are irritable about the crowds at our favourite swimming spot on Ruby Lake, remembering when it was just a tiny clearing among hardhack and cedars, the smell of wild mint pungent as we spread out our towels.

Farther afield, the subdivisions that cover Broadmead meadows near Victoria or the fields where I rode my horse along West Saanich Road in the years of my girlhood hang across my vision like unwelcome curtains. I can just see through them, a trick of the eye, to the dry grass where I ate my lunch while my horse waited, the smell of him still in my nostrils as I write this. In youth, we rolled our eyes as the old people we knew (some of them younger than we are now) talked about their earlier years and lamented the changes everywhere they looked. Where did time go, they asked, and our imaginations were too green to even think that we might one day remember everything we hardly noticed as we went about our lives.

And yet. And yet. I never expected to feel such physical loss as when I stand in the centre of my own still world and remember the past, which is almost always a landscape. Which is almost always what happened in a landscape. Camping trips with my brothers at Bamberton Beach, Englishman River, the shores of lakes where our father fished for our breakfast in his old jackets and our mother laid our Melmac plates on a picnic table and poured herself a cup of coffee from a battered aluminum percolator with a glass knob on top. Always there were trees which were gateposts to another world. The ponderosa pines I watched for in the Fraser Canyon, the spreading oaks on Quadra Street as I drove to my university classes or outside the window of the apartment I lived in on Fort Street, newly returned from a year in Ireland and waiting for the rest of my life to begin. Lying in my bed at night while car headlights illuminated the walls briefly, I'd dream my way back to Connemara, hearing the rasp of tough willow against the window of my cottage.

A section of Irish hedgerow
I'd cross over from Inishturbot by curragh to Eyrephort Strand and then walk up to the Sky Road where I might get a ride to Clifden if I was lucky. If not, I walked the eleven kilometres. Sometimes I borrowed a bicycle from the farmer whose cows grazed in the fields that ended at the sea. Either way, the road that led up to the Sky Road was narrow, a leafy tunnel through fuchsia, hawthorn, branches of black sloes hanging heavy from their stems, brambles, and gorse blooming in almost every month. I never knew all the birds that sang, or didn't, in the dense lattice of twigs and greenery but sometimes I'd see a nest with a blue tit hovering, or I'd hear the flute notes of a blackbird. Spiders, butterflies, bees humming in the primroses of early summer, and once I glimpsed a badger emerging from a gap where the hedge met a stone

wall. Cattle beyond the hedgerow grazed in sour fields while soft rain slicked their hides.

There weren't many large trees. Plantings of pines and yew near the farmyard of the bachelor who gave me rides a few times and who was handsome as sin but also rumoured to be dangerous. A few alders in the damp area where a seasonal stream came off the hills, the stunted willow by my bedroom window. I missed the dense forests of my native British Columbia raincoast during that year, though now I sometimes dream of walking up through that tunnel, fresh in spring or dust-worn in August, listening for birds, plucking a stem of fuchsia to tuck into my hat. Thirty-five years have passed, and still I remember white campion, dead-nettle, meadowsweet, and bryony lacing up into the sallies, and how I once dug up a small primrose to take back to my cottage where it bloomed in a blue teacup on the windowsill.

Always trees, with their leafy shade on a summer afternoon or fringes of needles to filter light and provide the scent of balsam to carry on my hands. They stand for hundreds of years on rocky outcrops, reaching deep into soil to anchor them until eternity or until a great unpredictable wind topples them and forever after is remembered as the wind that brought down the big fir or the hemlock by the bend in the driveway. (Do you remember?) Birds collect in their steadfast branches, climb their trunks (See the red-breasted nuthatch as I write this, racing up that cedar?), and the smaller ones nest in the holes left by bigger ones drilling for insects and sap. A quick whirl of chickadees comes to buzz and dart as I fill their feeder. I hold my hand out for a few moments only and one of them perches on my index finger, the lightest possible weight, pecking sunflower seeds from my palm. I know I am too impatient to wait much longer and the one brave chickadee doesn't return but scolds and agitates with the rest of them. Never mind. When I am old, I will stand for hours and let them choose from the seeds I carry. I'll stand so still they'll think I'm rooted.

Standing Dead

> How curious it would be to die and then remain standing for another century or two. To enjoy "dead verticality." If humans could do it we would hear news like, "Henry David Thoreau finally toppled over."
> — Gary Snyder, "Ancient Forests of the Far West" in
> *The Practice of the Wild*

There are many holes in the standing dead cedar on the curve of our driveway near the top. I've always wondered which birds nest there — red-breasted sapsuckers? Chestnut-backed chickadees? It's too far from the house for me to keep an eye on the comings and goings of the winged couples in spring, but I expect such perfect real estate is in high demand. Some birds no doubt return year after year. Not wanting to disturb their privacy, I instead imagine reaching into one of the high cavities and touching eggs among the hair-and-moss nests within the crumbling heartwood.

For years I loved to see another huge dead cedar near my home, part of a route we walked regularly with our wonderful dog Lily. I thought of all that tree had seen over the centuries on its slope above Sakinaw Lake. The water itself, alive with loons and trout. Canoes of the First Nations people, some of the paddlers painting messages on a tiny rocky island not far away. House-building by the people who are now our neighbours, then gardens blossoming in the mild air. Fireworks in winter. It was a perfect tree for a bear to stretch to, to sharpen its claws on the silver wood. For ants to enter, and colonize. For birds to perch on and nest in, for eagles to watch for merganser chicks and snap them up, one by one, on early summer mornings.

One winter afternoon, thirteen years ago, not long after a storm, we were walking the trail towards the dead tree when Lily stopped in her tracks, her ears alert. We expected to see an animal ahead but instead, the big tree lay across the trail. It had fallen in the storm, too far from our house to hear the crash. (If a tree falls and no one is listening…?) Horizontal, it came up to my chest, heart-high, and the bush adjacent to the trail was very dense and impenetrable. We certainly could have climbed, or detoured, but what about Lily, who was twelve and very arthritic, her hips creaky?

Finding long sturdy pieces of bark that had come away from the ancient trunk, we made a ramp for her and helped her climb to the top; then I held her while John scrambled to the other side and arranged the bark for her descent. It took some time. She wasn't happy about having to walk the plank, particularly as her eyes were cloudy with cataracts, and it was another moment — there are many when one is in the presence of an aging animal — when I realized that animals apprehend and navigate the world very differently from us. In youth this is true, and in age.

We didn't walk that trail again until the fallen tree had been cut into lengths by the neighbour who owned that piece of land, some lengths ending up milled into lumber, for the heart was clean and usable. The smell beside the area where the portable mill had been was heady — incense, spice, richly arboreal.

Not long after the tree toppled — we called it the fallen warrior — a dear friend died. He was a writer and his legacy is a shelf of books in my study. I look at them. "Dead verticality" or "dead, reclining," depending on how I arrange the volumes on my pine shelves. His name, Charles Lillard, on the thin spines, his dates bracketed in my memory.

And within a year, Lily died too. She has no literary legacy, nor offspring (she was spayed when we brought her home), no "dead verticality." But I do keep the bone of her pelvis on my desk. This is not as macabre as it sounds (or maybe it is). We had nowhere to dig a deep enough grave for her body so we dug out salal from beneath a big cedar on rough ground and laid her body among the roots and moss. We covered her with quantities of cedar boughs, moss, and slabs of bark we'd found near the tree. After two years, I reached into the area and pulled out a section of her skeleton which proved to be her pelvic girdle. It was quite clean but I put it into a bucket of water with a bit of bleach and let it sit for a week, then dried and aired it before bringing it into my study. I keep the long wires from the various accessories plugged into my computer — the mouse, the wire providing my Internet connection, the printer — coiled and nested in the open area near where her sacrum had gathered the fused bones of her vertebrae.

As a girl, I'd fractured my own pelvis when my horse fell on me after rearing on a hill in wind. In Lily's beautiful ivory bones, I see my connection to her, to my children (whom she loved and guarded like a mother), to the threads of life and death that hang close enough in our lives to touch at any moment.

As for Charles, I have his poems.

> The stars sang in the twilit garden;
> morning was moonlight,
> raspberries, wine clear as the wind and cold.[1]

And all along the length of that remembered tree, sawdust fell from the cavities where birds had nested in springs older than memory, deeper than love.

Go back, I tell myself. Go farther back, to the origins of the bike that took you to the very edges of the known world of Fairfield, along Dallas Road where the missionary sat among his spiders and plundered artefacts, past Clover Point. Dream your way back to the time when your brother came up from the swamp with your rocking horse over his shoulder, a beloved toy that had disappeared from the porch a year or two earlier, and which he found while out hunting frogs with his friends. It could not be restored, but you never forgot its return across the fields, even as you were yearning for something else to ride.

Juglans spp.
It was the autumn I was five years old and we lived in Matsqui, near the radar base where my father worked. Ten identical houses in a row provided housing for the families of the men at the base. There was a narrow boulevard running

in front of the houses and a grassed slope led up to the main road. Small trees were planted where the grass met the boulevard.

A neighbour's child, a year older than me, had received a bicycle for her birthday that fall. There was something so bold about the way that girl pedalled her bike back and forth along the boulevard and I was so jealous I wouldn't come out to play with her and the other children of the row. In later years, my parents confessed they hadn't intended to give me a bike of my own that Christmas, but there didn't seem to be any other solution to my anguish. This surprises me now, because it would never have occurred to me that my actions would ever warrant such results.

But there was a bike, a small blue bicycle, under the tree that Christmas, and I was ecstatic. I didn't know how to ride it. If I hadn't spent the weeks leading up to the holiday in my bedroom, green with envy, I'm sure the child next door would have taught me to ride her bicycle.

I don't know what gave me the notion that I should take my new bike to the top of the grassy slope, once I'd learned how to balance myself briefly on the seat, toes to the ground. Maybe other children were present and suggested it, but in my memory, I was alone on that rise—which was probably very minimal in any case (I've revisited hills that I remember as treacherous in my childhood and realized that their mild incline was such that I didn't even have to shift gears to drive up them). After balancing for a few moments at the top—in triumph—I let myself go and coasted down the hill.

It was thrilling in the extreme—the speed, the freedom of rushing through cold air, the way I felt as bold as that neighbour girl a month or so earlier. But then I realized I would have to stop or else I'd crash into the house on the opposite side of the boulevard. In panic, I steered my new treasure into a tree at the bottom of the slope and collapsed as the front wheel hit the trunk hard enough to bend the bike frame slightly.

I may have hurt myself—I don't remember this—and I know that my father was not happy to have to straighten the frame of my new bike in his workshop so soon after its arrival into my life, but after all these years, I still remember the smell of the bruised bark of that tree. Walnuts! Somehow it had never occurred to me that nuts grew on trees, and that the young trees outside our house might be that very kind. On my hands and scraped legs, I crouched by the tree and pressed my face to the bark. Yes, walnuts.

We didn't have nuts very often, but at Christmas a bowl of them sat on our coffee table with a nutcracker and a metal pick to extract every morsel from the shells. Brazil nuts, filberts, almonds, pecans in their shiny deep red shells, and the walnuts, my favourites. Such luxury to sit on the rug and eat nut after nut while watching *Looney Tunes* on our television while the Christmas tree glittered in its corner, dressed in tinsel and lights.

In later years, I walked along Victoria's tree-lined streets, eating fruit from the ornamental plum and cherry trees. Most of this fruit was sour, but the notion that a tree could provide bounty was magic to me. And in some of the backyards of the many houses we lived in over the years, there were apple trees — fruit following the sweet blossoms as regular as clockwork every year. Such a gracious gift to a child, boughs drooping with apples or small hard plums or best of all, the possibility of walnuts engendered by bruised bark on Christmas Day in 1960.

Acer macrophyllum
In fall, the samaras whirl to the ground: time to be grateful for fire, the woodshed neatly stacked with fir and bigleaf maple. Bringing in logs, I sometimes see areas of spalting within the chunks of maple I carry. This is a bacteria that causes veining in the wood, a kind of scribbling, like pen lines on paper. The bacteria can be introduced to felled maple, and cultured or managed for a time, to create beautiful patterns which woodworkers value. We have a cutting board in our kitchen made by a local craftsman, featuring strips of both spalted and clear-grained maple. When I clean and oil the board, I marvel at the intricate text in the wood we use to cut our bread. Like those beetles that wrote obituaries to the ponderosa pines near Kamloops, something lively is at work to leave its story intact for the future to read as loaves are sliced, fish boned or trimmed of their fins.

There are many maples in our woods, some of them mossy with age... The bigleaf maple is one of the glories of these western forests. Their honeyed flowers in spring are loud with bees and the fallen ones are dense with small flies. In summer they provide canopies that keep temperatures a little more moderate than surrounding areas. The edges of Sakinaw Lake Road are thick with their humus in fall, habitat for the rough-skinned newts we've found in the decaying leaves. And in winter we see the beauty of their bare architecture, the revelation that their trunks and branches host fern colonies and even smaller trees growing from deep mossy pockets established in clefts and crotches.

A whole area of study, called canopy biology, concentrates on the ecology of these arboreal communities of epiphytes, hemiepiphytes, climbers, insects, amphibians, etc. Remembering how we climbed maples as children, spread ourselves along their mossy boughs, I wonder if I could climb again to enter that upper realm, the flowers just opening and yellow-rumped warblers trilling among them.

The past is almost always a landscape — what happened in groves of trees, those suede hills trembling with aspens on the road out of Lillooet or north of Merritt, the ancient cedars fertilized with centuries of salmon carcasses at Goldstream Park. I remember that love song inspired by a plane tree as Xerxes led his army to Greece and the silvery olive trees as I sat by the window of a bus hurtling across Crete, wondering at the next chapter of my life. Or earlier, the scent of walnuts, the beautiful shape of Garry oaks in winter, the ones on the hill by the Protestant Orphanage in Victoria settling into darkness like a herd of elk, the night and everything else tangled in their antlers. Their afterimage stays with me. Some nights, it's the last thing I see before sleep.

The world changes, and never changes. In the time that is always mine, I walk home from school down Haliburton, across Pat Bay Highway, along Elk Lake Drive where slowly a field becomes a hotel, a few houses disappear, and the trees I knew as a girl recede into the hazy line that is no longer a horizon but a dream. Fences articulate such small space! The area that was once the

Yale estate, known as Colquitz Farm, has long since been subdivided. Still, a few old trees, a deodar cedar and a tulip tree, remember the old days on their corner where traffic races past, not far from a storefront built by grandsons of Hannah and Richard Maynard. Those trees remember the children and grandchildren of Hudson's Bay Company Chief Factor James Murray Yale on their way to the schoolhouse at Royal Oak, and Mr. Kinnaird going to choir practice at the Wilkinson Road Methodist Church. The imprint of history is everywhere I look, though I am late in realizing it. The Quick farmhouse on its hill above Wilkinson Road, daffodils running down to the road each March, where cattle once grazed and rested—so often I rode my horse along that road and never thought about the people who had cleared the land and planted the lilacs and other ornamentals. Who sold milk to the community and who campaigned to have the municipal and community halls built in their neighbourhood and who were active in both organizations. Later I walked down into Quicks Bottom, now a wildlife sanctuary behind the original farm, where blackbirds whistled in the canary grass and blue camas covered the ground under Garry oaks and black hawthorns. I bent to look at a bee in an open fawn lily, suddenly seeing low mottled leaves everywhere in the filtered light. A rabbit watched, bemused, from the other side of the fence, and a small snake dozed in a pocket of sunlight.

Underplantings
Through the allée on the Klein Lake trail, where masses of *Viola sempervirens* bloom in early to late March, clusters of the heart-shaped leaves cradling the sweet yellow violets. Where thick mats of *Linnaea borealis*, favourite of its Swedish namesake (and father of the binomial system of classification), bloom in May and June, the shady path a distillation of the perfume rising from each nodding twinned flower. "Ahh," we say, falling to our hands and knees to plunge our faces into the tiny groves, "just like almond extract!" Where *Trientalis latifolia*, the western starflowers, are carried aloft on their thin stems to form airy constellations, anchored in earth by little potatoes.

And then the lilies: *Lilium columbianum* in the grass at the edge of the path, many blooms to a stalk, smelling like mandarin oranges; the fawn lilies (*Erythronium oregonum*) above their dappled leaves in the shade of arbutus and shore pine at Francis Point and Quicks Bottom every Easter, late or early, so many of them I want to weep for their beauty; and once, driving back from Campbell River, we decided to walk at the Oyster River estuary and saw hundreds of *E. revolutum*, a deep pink form. On the slope where we stop to sit in sunlight, the eerie *Zygadenus venenosus*, or death camas, in the grass already drying in late May; a few *Fritillaria lanceolata* are hidden among them, speckled like birds' eggs. It looks like a perfect place for blue camas but I've never seen it here, though in Victoria, drifts of *Camassia quamash* cover the meadows of Beacon Hill Park, the roadsides in Metchosin. I remember driving to see Forrest at Lester Pearson College of the Pacific and stopping the car because the camas was in full bloom, shades from pale to deep blue, as common as grass. In the graveyard of the Church of Saint Mary the Virgin, fawn lilies covered the ground under the Garry oaks.

As a child, near Clover Point, I sat on rocks by the sea and idly ate grass stems, as a child does, and was surprised that some tasted of onion. I didn't know then about *Allium cernuum*, the nodding onion, but thought of it later, in Ireland, when I found wild garlic in the ditches under the hedges or other damp places, its flowers as lovely as any spring bulb. I used to pick it and keep it fresh in a glass of water in my island kitchen because I was so poor I often couldn't afford to buy food and relied on nettles, mussels, and the buckets of potatoes my neighbour brought from his garden. Boiled together and flavoured with snippings of wild garlic, they were my dinners for weeks at a time.

How to watch the ground at my feet, the footnotes, and walk at the same time? What have I missed, trying to protect my footing on the mountain trails? Clubmoss sprawling across the gravel, wild ginger, enchanter's nightshade on the shady path by the waterfall, rattlesnake plantain rising from its rosettes on a rubbly turn where the trail switchbacks to the summit, mimulus and mouse-eared chickweed, long strands of honeysuckle straggling through the ocean spray. What have I missed, waiting for the first cerise salmonberry blossom, as early as February 23 one year,[2] or brushing away debris to find the tiny prince's pines under the firs on the Hallowell loop trail? Whole lives have passed me by while I bend to uncover a trillium, a clump of *Viola adunca* in our orchard where the ground is like the carpet of wildflowers in Botticelli's *Primavera*, violets, self-heal, vetch, everlasting pea, and Columbia lilies in the soft grass. It is always spring and I am always young — though right now it is winter and I am newly fifty-five, dreaming that I am filling my hands with beauty.

Always spring; or always winter — the scent of evergreens, scarlet hips on the climbing roses, grouse feasting on small scabby crabapples, a long fluid line of elk running up the mountain when we walk on the high Malaspina trail.

Abies grandis
Walking on that high trail in the fall of 2008, we noticed a single grand fir growing behind a huge stump — this is in the area kept clear because of hydro pylons overhead — and decided it would make a wonderful Christmas tree. Perhaps two metres tall, it was full and symmetrical, its branches dense. Every week we walked up there with our dog Tiger and every week we'd check the beautiful tree, crush some needles in our hands for the balsam odour, and reassure ourselves that no one else would see the tree there hidden in huckleberry and salal. In truth, the area is a good place for tree hunters as small firs grow in the open under the hydro lines and free permits are available for those who want to cut their own Christmas trees. Most of these trees are *Pseudotsuga menziesii*, the Douglas fir, not true firs at all, but the source of fine timber on the coast. There are also some pines up under the hydro lines, and hemlocks. But our tree was well-concealed.

We always decorate our tree on the afternoon of the day before Christmas. Years ago this was a way to fill the day with practical activity when we had three small children eager for what would come the next morning. In those years we usually cut a tree on our property. We have more than three hectares and the expedition to find exactly the right tree could take an hour or two, John armed with the folding pruning saw and accompanied by helpers. They'd look for a tree growing very near another, reasoning that the thinning was good for our woods. The trees were often bare of branches on one side but that made it possible to place them against a wall where they'd take up less room and no one noticed that the back was a little sparse. I'd stay behind and heat cider on the stove, festive with slices of orange pierced with cloves, sticks of cinnamon broken into the pot. Out would come the boxes of ornaments stored in a dark cupboard for the rest of the year and we'd drink the cider while decorating the tree. Each year the delight, as old and new favourites emerged from the box — the Paddington bear, stained glass stars created by our friend June, each one more exquisite than the last, the paper lanterns sent to John's family from his grandmother in Suffolk their first year in Canada, the angels and Santas made at school and decorated with macaroni. Some years there would be family members to help: my parents often came; John's mother and father (separated and living in different cities so a bit of juggling was required to invite one and not the other); friends or relatives who were free that year to join us.

Even though our children have grown up and gone away, they still come home for the holiday and we still save the tree cutting and decorating for the afternoon of Christmas Eve. But it snowed the week before Christmas in 2008, heavy falls, and the roads iced up to the extent that our highways maintenance workers couldn't keep up with the clearing. It was the year Forrest successfully defended his PhD dissertation and treated himself to a train trip home on the Empire Builder from Chicago to Seattle. Heavy snow in Ontario and the American Midwest made that both a disaster and an adventure, and he arrived two days later than anticipated, sleep-deprived, at night, on the bus, with us at the side of the frozen highway armed with flashlights to show the driver where to stop as every familiar post or driveway was covered by about a metre of snow. Our plan to drive up the mountain on Christmas Eve to cut the grand fir had to be quietly abandoned and a tree was cut close to the house, a wispy inelegant Douglas fir that I insisted was "lyrical" when it was laden with its lights and ornaments. Anyway, we said, the grand fir would be there next Christmas, even nicer for the wait.

Several days before Christmas Eve, 2009, John and Brendan went up for the tree — we'd kept an eye on it all the next year, watching it grow even more beautiful, and more hidden as the wiry huckleberry sheltered it — reasoning it was better to bring it home a little early in the event snow fell or something else happened to make it impossible. It could wait for a few days in the woodshed, its trunk in a bucket of water.

They were gone a long time. When they returned, they were the bearers of bad news. Someone else had taken our tree! How was that possible, I wondered, remembering how hard it was to see it behind its fringe of brush. But it was true. They'd driven up the hydro road, over the washed-out area, to the big mossy stump and found only a smaller stump behind it, sticky with sap.

But they were also the bearers of a lovely bushy Douglas fir they'd found much farther along the hydro road, up past the creeks which they'd had to ford in the Honda Element, and beyond the area where the flowering currants are alive with hummingbirds in spring. And it waited in the bucket in the woodshed, upright, a dense vertical apparition which startled me over and over first thing in the morning when I'd look out the kitchen window, wondering who on earth was standing outside at that hour.

I had imagined our house filled with the scent of grand fir at Christmas, the resiny blisters on the bark oozing their fragrant balsam. I'd imagined a room perfumed with our own equivalent of frankincense and myrrh, those ancient offerings from distant trees, though in truth the Douglas fir smelled wonderful, a distillation of forest and damp green, a familiar balm to come down to on Christmas morning with the old carols calling us to rejoice, be merry.

Sometimes I wake from dreams of places I knew as a girl, dreams so vivid and natural, that I ache to return. A warm cleft of rock at East Sooke Park or Sandcut Beach, my face bathed in spruce-scented wind. The cluster of fawn lilies just beyond my bedroom in the charmed house at Yarrow Point — their sweetness in April an unexpected pleasure as I went out, sleepless, in moonlight, by the green door. The groves of cedars along the Goldstream River: in a dream I am walking there still, a long-dead dog racing ahead for the joy of water.

"Live in the layers, / Not on the litter," Stanley Kunitz advised in a late beautiful poem,[3] and I have taken this to heart. Trees have their inner and outer layers of bark, and a layer of cambium, then sapwood and heartwood. So much is contained there! A record of years, weather, visitations by insects and fungi, seasons of drought and abundant rain.

What happened in a grove of trees? In the first place, a life, my life, accumulated there. I walk through, remembering, stopping at each tree — pines, cedars, firs, the unlikely olives and planes, Garry oaks, live oaks, the beeches of my lost grandfather's Bukovina (and the newly planted copper beech, caged in deer-proof wire, in memory of my father, waiting for its benediction of ash), arbutus on an island I sailed to as a young woman, the trembling aspens passed on my way to a wedding, and the arboretum of rare or cherished plantings. "The best aid to clearness of memory consists in orderly arrangement," said Cicero in his *De Oratore*,[4] and I despair of such orderliness. Everything comes to me in such splendour and chaos. Am I right in remembering the owner of the Rolls-Royce as the delivery man at the pharmacy where I worked? Or that I drank raki on a quay on Crete in the morning while my hands bled from rough ropes, that I ever wept (in the second place) on the side of the highway while listening to David Daniels sing Handel?

Every day I walk out of the house I built with my husband into a landscape of trees, some older than the country, some planted more than half a century ago to replace others felled by loggers who cut the local forests to the ground. They are still cutting trees, a team working on the mountain above the Malaspina substation where we walk most weeks. I hate to see the limbed giants lying in piles to be loaded onto trucks and taken to mills, pyramids of slash heaped to be burned. I wonder about the birds and animals that made their homes in those forests, though this operation worked through the winter, a dormant season. New regulations require that the cut-blocks are smaller, there is less waste, and in any case, this is an important part of our coastal economy. In fact, I recognize some of the young men working up there. They were classmates of my children. They've remained in the community, bought houses, and some of them have children of their own.

I walk out into cedars, red alders, Douglas firs, down our long gravel driveway carved out of western hemlock and salal, past salmonberry bushes, their buds barely containing the brilliant petals ready to unfurl. Some days, new scats, dense with tiny bones, tell me coyotes have come up to the house while we're sleeping and in late summer, bears climb into the crabapple to feast on its scabby fruit. I have my eye on a little patch of prince's pine, hoping I won't miss its brief season, the nodding pink flowers worth kneeling to the ground for. Ravens tumble on the thermals, klooking and tocking, and some days I talk back to them, giving them what I hope is a report on the state of things below. I wish I could sing to them but they've demonstrated before that they are the divas, their watery arias performed with high voice, or low.

In his splendid *Sylva*, John Evelyn wrote,

> *Thuya*; by some call'd *arbor vitae*, (brought us from Canada,) is an hardy green all the Winter, (though a little tarnish'd in very sharp weather)...most delights in the shade...The leaf being bruised between the fingers, emits a powerful scent not easily conquer'd.[5]

The tree of life, its wood supporting our aging wisteria, also starts the fire in the kitchen each day in its incarnation as kindling. Most mornings of my life, I smell its smoke, a complex incense, summoning Goldstream Park, the spicy scent of fresh-sawn boards, the boughs I cut to wreathe the front door each Christmas. In the *Tusculan Disputations*, written in retirement, Cicero meditates on memory: he wonders at the power that inspires us to invent, to wander, discover, create shelter as protection from wild animals, learn astronomy, create music, poetry, and philosophy. "For what is the memory of things and words?" he asks. "Assuredly nothing can be apprehended even in God of greater value than this..."[6] Cicero concludes that our ability to remember is a proof of the divinity of our souls.

As a child and young woman, I explored a place set in my mind like a petroglyph. The images come to me in my daily life—the plants and trees of Fairfield, of Royal Oak, the shimmering olive groves of Crete, where, echoing Ovid's "The forest's a house, the leaves a bed,"[7] I stopped with Agamemnon to lie in the myrtle. I am sustained by trees. My life unfolds among the shade of the coastal rainforests where my house is anchored like a deep-drinking arbutus tree, eager for bedrock. It's a nest box, waiting for swallows, for children far-flung and missed. I've lost track of their departures but never their returns.

It rains a lot here. A tree can grow to an immensity undreamed of in other parts of the world. On the Klein Lake trail where we walk every week, huge firs, hundreds of years old, are draped with common witch's hair and speckled horsehair, browsed by deer in winter. Every morning the sun rises above Mount Hallowell to the east of us and every evening it sets to the west, behind Texada Island and the Strait of Georgia beyond. Dante wrote of the dark wood, in the middle of his life: *To tell about those woods is hard—so tangled and rough...* Yet, when the rain stops, sunlight comes through the trees so clear and true that the damp world shines.

Acknowledgements

I would like to acknowledge the following journals in which versions of some of these essays first appeared:

Lake ("*Quercus garryana*: Fire")
Dandelion ("*Pinus ponderosa*: A Serious Waltz")
Memewar ("*Platanus orientalis*: Raven Libretto")
Brno Studies in English ("*Quercus virginiana*: Degrees of Separation")
Contrary ("*Olea euroropaea*: Young Woman with Eros on her Shoulder")
The New Quarterly ("*Arbutus menziesii*: Makeup Secrets of the Byzantine Madonnas," winner of the Edna Staebler Personal Essay Contest)
Cerise ("*Thuja plicata*: Nest Boxes")
Lived Experience Number 11 (*Populus tremuloides*: Cariboo Wedding")

"*Platanus orientalis*: Raven Libretto" is dedicated to the memory of Floyd St. Clair.

Many people provided information, suggestions, encouragement, and inspiration during the writing of this book. I thank them all but particularly my husband John Pass and our children Forrest, Brendan, and Angelica Pass. They animate the pages as they animate my life, with patience and love.

I'm grateful for the gracious and intelligent editorial guidance of Akoulina Connell, the careful and astute eye of Paula Sarson, as well as the enthusiastic support of everyone at Goose Lane Editions.

I would also like to thank the following individuals and companies for their generosity in granting permission to quote passages from their work. Every effort has been made to secure permission for excerpts reproduced in this book. I regret any inadvertent omissions.

11, 229 Excerpt from "Canto 1" from *The Inferno of Dante: A New Verse Translation* by Robert Pinsky. Translation copyright © 1994 by Robert Pinsky. Published by Farrar, Straus, & Giroux, LLC and Orion Publishing Group. Reprinted by permission.

18 Quotation from "Time to Burn" by Nancy J. Turner in *Indians, Fire, and the Land in the Pacific Northwest*, edited by Robert Boyd, copyright © 1999. Reprinted by permission of Oregon State University Press.

20, 21 Quotation from Aonghas MacNeacail. Reprinted by permission of the author.

27 NCB77 courtesy of Royal BC Museum, BC Archives.

30-31 Quotation from Ted Lea. Reprinted by permission of Ted Lea.

33, 131 Quotation from *Natural History: A Selection* by Pliny the Elder, translated with an introduction and notes by John F. Healy (Penguin Classics, 1991). Copyright © 1991 by John F. Healy. Reprinted by permission.

36 Quotation from Royal Botanic Gardens, Kew Web site. Reprinted by permission of Royal Botanic Gardens, Kew.

47 Quotation from *Trees of Greater Victoria* by G.D. Chaster, D.W. Ross, and W.H. Warren © 1988. Reprinted by permission of the Heritage Tree/Book Society of Greater Victoria.

52 Quotation from *Good Intentions Gone Awry: Emma Crosby and the Methodist Mission on the Northwest Coast* by Jan Hare and Jean Barman copyright © 2006 by University of British Columbia Press. All rights reserved.

52 Quotation from *Captured Heritage: The Scramble for Northwest Coast Artifacts* by Douglas Cole copyright © 1995. Reprinted by permission of University of Oklahoma Press via the Copyright Clearance Center.

54-55 Quotation from Ira Jacknis. Reprinted by permission of the author.

56 BCPM corr., GR111 Box 8, File 39, Dec. 4, 1953 Wilson Duff to Richard Conn Reprinted by permission of Royal BC Museum, BC Archives.

61, 70, 79, 80 Odysseas Elytis excerpts from "Anoint the Ariston" and "As Endymion" from *Eros, Eros, Eros Selected and Last Poems*, translated by Olga Broumas. Translation copyright © 1998 by Olga Broumas. Reprinted by permission of The Permissions Company Inc. on behalf of Copper Canyon Press, www.coppercanyonpress.org.

62 Extract from part 9 of "Mythistorema" from *George Seferis: Complete Poems*, translated by Edmund Keeley and Philip Sherrard. Published by Anvil Press Poetry in 1995. / Keeley, Edmund; *George Seferis*, copyright © 1967 Princeton University Press, 1995 renewed PUP/1995 revised edition. Reprinted by permission of Princeton University Press.

67, 76 Quotations from Stephen G. Miller's *Arete: Greek Sports from Ancient Sources*. Reprinted by permission of the University of California Press.

73 Quotation from *The Colossus of Maroussi* by Henry Miller, copyright © 1941 by Henry Miller. Reprinted by permission of New Directions Publishing.

119, 129 Quotations from Theocritus. Reprinted by permission of Anthony Holden.
128-129 Quotations from *Orfeo ed Euridice*. Translation by Andrew Huth. Reprinted by permission of Decca Classics.
158 Excerpt from "Book 13" from *The Odyssey* by Homer, translated by Robert Fitzgerald. Copyright © 1961, 1963 by Robert Fitzgerald. Copyright renewed 1989 by Benedict R.C. Fitzgerald, on behalf of the Fitzgerald children. Reprinted by permission of Farrar, Straus, & Giroux, LLC, and the Estate of Robert Fitzgerald.
164 Quotation from Alice Glanville's *Schools of the Boundary: 1891-1991*. Reprinted by permission of Sonotek Publishing.
194 Quotation from Patrick Kavanagh's poem "On Raglan Road." Reprinted by kind permission of the Trustees of the Estate of the late Katherine B. Kavanagh through the Jonathan Williams Literary Agency.
197 Quotation from *Illustrated Flora of British Columbia, Vol. 5, Dicotyledons (Salicaceae through Zygophyllaceae) and Pteridophytes*, G.W. Douglas, D. Meidinger, and J. Pojar, eds. B.C. Ministry of Environment, Lands and Parks, and B.C. Ministry of Forests, Victoria, 2000, 389 pp. Reprinted by permission of the Ministries.
200, 203 Quotations from *Plants of Southern Interior British Columbia and Inland Northwest*, Roberta Parish, Ray Coupé, and Dennis Lloyd. Reprinted by permission of Lone Pine Publishing.
216 Quotation from *The Practice of the Wild* copyright © 1990 by Gary Snyder. Reprinted by permission of Counterpoint.
218 Quotation from "Closing Down Kah Shakes Creek" by Charles Lillard. Reprinted by permission of Rhonda Lillard.
227 Quotation from *The Collected Poems* copyright © 2000 by Stanley Kunitz. Reprinted by permission of W.W. Norton & Company.
227 Quotation from Cicero. Reprinted by permission of the publishers and the Trustee of the Loeb Classical Library from *Cicero: Volume III, De Oratore*, Loeb Classical Library Volume 348, translated by E.W. Sutton and H. Rackham, p. 353, Cambridge, Mass.: Harvard University Press, Copyright © 1942 by the President and Fellows of Harvard College. Loeb Classical Library ® is a registered trademark of the President and Fellows of Harvard College.
228 Quotation from Cicero. Reprinted by permission of the publishers and the Trustees of the Loeb Classical Library from *Cicero: Volume XVIII*. Loeb Classical Library, Volume 141, translated by J.E. King, p. 65, Cambridge, Mass.: Harvard University Press, Copyright © 1927 by the President and Fellows of Harvard College. Loeb Classical Library ® is a registered trademark of the President and Fellows of Harvard College.
229 Quotation from Ovid's *Ars Amatoria*, translated by Angelica Pass. Reprinted by permission of the translator.

Endnotes

Prelude
i. Cicero's ideas of mnemonic placement are distilled in Book ll of *De Oratore*, lines 350-360.
ii. Evelyn, *Sylva,* p. 56.

Quercus garryana: Fire
1. Jon Keeley has researched the effects of fire on seed germination in California in particular; Gavin Flematti's work has concentrated on fire and native plant ecology in Australia. I've included some of their articles in the bibliography.
2. Nancy J. Turner, "Time to Burn: Traditional Use of Fire to Enhance Resource Production by Aboriginal Peoples in British Columbia," in *Indians, Fire and the Land*, ed. Robert Boyd, p. 196. This essay led me, via its notes on cited materials, to the source of the quoted material: correspondence between James Douglas and James Hargrave, in *The Hargrave Correspondence, 1821-43*, ed. G.P. de T (George Parkin de Twenebroker) Glazebrook, recommended reading for the window the correspondence provides onto the workings of the HBC during this seminal period in its influence.
3. Captain George Vancouver is quoted in Nancy Turner, "Time to Burn," p. 195: "I could not possibly believe any uncultivated country had even been discovered exhibiting so rich a picture...extensive spaces that wore the appearance of having been cleared by art."
4. Anecdotal material cited in Nancy Turner, "Time to Burn," p. 200.
5. Ibid. p. 194.
6. Information on the etymology of "oak" comes from several sources, notably my household's hefty *Compact Edition of the Oxford Dictionary, in two volumes* (Oxford University Press, 1971); Bill Casselman's *Canadian Garden Words*; Peter Wyse-Jackson's *Irish Trees and Shrubs*; and Peter Harbison's *Ancient Irish Monuments*.

7. This is discussed by Andy Coghlan in "Sensitive Flower," *New Scientist*, September 26, 1998, pp. 23-24.
8. I've read a fair bit about the Ogham (or Ogam) alphabet and know that there are many theories about its origin. Perhaps the most compelling and least fanciful is presented by Damian McManus in the monograph included in the bibliography; the *Archaeological Survey* of Judith Cuppage is also based on solid research. The holed ogham stone at Kilmalkedar predates the seventh-century Christian monastery and is probably pre-Christian in origin. Many ogham stones can be found on the Dingle Peninsula.
9. Poem by Aonghas MacNeacail. Used by kind permission of the author. (I first read this poem in the Times Educational Supplement online. Web address given in bibliography.)
10. Rhapsodes or rhapsodists were professional reciters of poetry. They would perform at festivals and games, often in competition for prizes. The Homeric Hymns are generally believed to have been composed by a number of professional rhapsodes, almost certainly not by Homer himself but in the same tradition, though a little later — between the eighth to sixth centuries BCE. This excerpted passage is from Hesiod, *Homeric Hymns and Homerica*, 127.
11. Ibid.
12. The material on the British Columbia Protestant Orphans' Home comes from various sources, among them http://web.uvic.ca/~cduncan/orphanshome/poh.html and Derek Pethick's *Summer of Promise, Victoria 1864-1914*. The distasteful term "half-caste" comes from a letter quoted at the Web site cited here, and I use it to demonstrate the attitude that prevailed in the city at that time. The letters between Flora Sinclair and Mary Cridge, NCB77, are used courtesy of the Royal BC Museum, BC Archives.
13. Information on historical distribution of Garry oaks on south Vancouver Island came in part from personal recollection as well as informal discussions with many people. I looked at many archival photographs of Saanich Peninsula. I also consulted the Web site of the Garry Oak Ecosystem Recovery Team — http://www.goert.ca — and am grateful to Ted Lea, a retired vegetation ecologist with the Ecosystems Branch of the BC Ministry of Environment, who sent me the map he helped to prepare. I hung the map near my desk and spent many hours musing about the lost landscapes of my childhood.
14. Personal email correspondence from 2008. Used with permission.
15. The story of the three broom seeds planted at Sooke by early settler and HBC surveyor Walter Colquhoun Grant sounds apocryphal but occurs in so many sources that it must contain at least a germ of truth.

Quercus virginiana: Degrees of Separation
1. Douglas's report, Fort Vancouver, HBC, July 12, 1842, in *Botanical Electronic News*, #226, July 2, 1999.
2. Ibid.
3. Janis Ringuette, *History of Beacon Hill Park*, Appendix C: Dallas Road Waterfront, 2009, www.beaconhillparkhistory.org/.
4. Archie H. Wills, "Booze and Bullets Flew in the Days of Cliff House," in *Daily Colonist*, Feb. 28, 1971, p. 13. Noted in Ringuette at note 3.
5. "The Provincial Museum had undergone two name changes, from the Provincial Museum of Natural History and Anthropology, to the British Columbia Provincial Museum to the Royal British Columbia Museum" (http://www.royalbcmuseum.bc.ca/Natural_History/Birds.aspx?id=723). In this chapter, I've tried to use the name appropriate to the period I'm writing about and, of course, there are considerable shifts in time from Newcombe's collecting years to my childhood in the early 1960s to the present.
6. The Kwakwaka'wakw people are also known as Kwakiutl or Kwaguilth — though more properly "Kwakiutl" refers specifically to those people belonging to the Fort Rupert Band. Traditional Kwakwaka'wakw territory is bounded in the north by Smith Inlet, the south by Cape Mudge, the west by Quatsino Sound, and in the east by Knight Inlet. The Lekwungen people who occupied what is now Victoria are members of the Coast Salish linguistic group.
7. Chaster, Ross, and Warren, *Trees of Greater Victoria: A Heritage*, p. 43.
8. *The British Colonist*, June 15, 1872, at britishcolonist.ca/.
9. Jan Hare and Jean Barman, *Good Intentions Gone Awry*, p. 85.
10. Douglas Cole, *Captured Heritage*, pp. 22-23.
11. Ibid. p. 191.
12. Ibid. p. 85.
13. Wilson Duff, *Thunderbird Park*, p. 20.
14. Phil Nuytten, *The Totem Carvers*, p. 86.
15. Ira Jacknis, "Authenticity and the Mungo Martin House, Victoria, BC: Visual and Verbal Sources," in *Arctic Anthropology*, p. 7.
16. This term comes up in both anthropology and cultural studies to denote the kind of collecting mentality that was so prominent in the early twentieth century when ethnographers felt compelled to "save" whatever aspects of material culture they could find, preferably the oldest and most authentic, in the belief that colonial influences would result in the disappearance of indigenous cultures. I'm grateful to Dr. Michelle Hamilton, Assistant Professor and Director of Public History at the University of Western Ontario, and the author of *Collections and Objections: Aboriginal Material Culture in Southern Ontario* (McGill-Queens's University Press,

2010), for clarifying the term itself as well as its historical context. Dr. Hamilton said, "Some say Jacob Gruber coined the term salvage ethnography/anthropology but I've found the term used earlier than his article, at least in the 1950s. Nevertheless, I usually use his article to footnote the concept as it is one of the only articulations of the concept. The article is: Gruber, Jacob W. "Ethnographic Salvage and the Shaping of Anthropology."*American Anthropologist* 72, 6 (1970): 1289-99. I could also suggest Marvin Harris' *The Rise of Anthropological Theory* which is a standard survey. As for the word 'paradigm,' I tend to avoid use of what historians consider to be post-modern terminology while anthropologists and others are more comfortable with it." From an email correspondence, March 2011, used with permission.
17. Wilson Duff to Richard Conn, 4 December 1953. BCPM Correspondence, GR 111, box 8, file 39, courtesy of Royal BC Museum, BC Archives.
18. Robin Ward, *Echoes of Empire*, p. 58.
19. *The British Colonist*, November 1, 1903, at britishcolonist.ca/.

Olea europaea: Young Woman With Eros On Her Shoulder
1. George Seferis, *Collected Poems*, p. 25, used with kind permission of the translator Edmund Keeley, Anvil Press Poetry, and Princeton University Press.
2. Ibid.
3. Julius Pollux was a second century AD Greek scholar and teacher. He's best known for his *Onomasticon*, a kind of thesaurus in ten volumes, mostly lost, though an abridged Latin version is extant. This passage occurs in Stephen G. Miller's *Arete: Greek Sports from Ancient Sources*, p. 125, and is used with Dr. Miller's and the University of California's generous permission.
4. Elytis, *Eros, Eros, Eros*, p. 72, used with kind permission of the translator Olga Broumas.
5. Henry Miller, *The Colossus of Maroussi*, p. 163.
6. Elytis, *Eros, Eros, Eros*, p. 157.

Thuja plicata: Nest Boxes
1. John Dowland's *First Book of Songs and Ayres* appeared in 1597 and contained arrangements for lute and voice.
2. Gaston Bachelard, *The Poetics of Space*, p. 48.
3. Ibid. p. 100.
4. Ibid.

Platanus orientalis: Raven Libretto
1. Although I own, and admire, the Landmark Herodotus (edited by Robert Strassler and beautifully translated by Andrea L. Purvis), I chose the nineteenth-century translation of George Rawlinson for this epigraph. To my ear and mind, it captures the capricious quality of Xerxes's infatuation with the plane tree on the banks of the Maeander: Herodotus, *The Histories,* p. 272.
2. "*Ombra mai fu*" is an aria from *Serses* (or *Xerxes*), a 1738 opera composed by Handel, libretto by Nicolò Minato.
3. David Daniels, *Operatic Arias*, Orchestra of the Age of Enlightenment, Sir Roger Norrington conducting, Virgin Veritas CD.
4. *Theodora,* Act 1, Scene 4. Composed by Handel, libretto by Thomas Morell. I've watched the extraordinary Glyndebourne Festival Opera production of this 1749 oratorio on DVD, directed by Peter Sellars, with a cast featuring Dawn Upshaw singing the role of Theodora, David Daniels singing Didymus, and Lorraine Hunt Lieberson singing Irene. Libretto available at http://opera.stanford.edu/iu/libretti/theodora.htm.
5. My daughter Angelica bought this translation of Virgil's *Georgics*—by Smith Palmer Bovie, published by University of Chicago Press, 1966—at a book sale at the University of Victoria. I was enchanted to discover that its previous owner, as inscribed on its overleaf, was the late Peter Smith who'd been my Classics professor at UVic in the mid-1970s. He was one of the most generous and erudite individuals I encountered in my university life. The poem excerpt comes from Book IV, pp. 145-147.
6. It's hard to disentangle fact and fiction from stories about Hippocrates, the fifth-century BCE Greek physician about whom little is known but much is suggested. But there is a plane tree on Kos (or Cos, presumed to be the island of his birth) associated with him and his teachings; this association seems to have the endorsement of the British medical establishment! A.S. Playfair, "Hippocratic Plane," in *Journal of the Royal College of General Practitioners,* June 21, 1971, pp. 367-368.
7. Occurrence of fossil plane leaves and seeds discussed by Dr. Bruce Cornet at http://www.sunstar-solutions.com/sunstar/Sayreville/Kfacies.htm.
8. Description of leaves from Maggie Campbell-Culver, *A Passion for Trees,* p. 169.
9. Edward Gibbon, *The History of the Decline and Fall of the Roman Empire,* p. 494.
10. Evelyn, *Sylva,* p. 216.
11. This epigraph and all the passages in this section are from Henry Purcell's opera *Dido and Aeneas,* libretto by Nahum Tate, composed sometime before 1688. I love the 1967 recording, conducted by Sir Charles Mackerras, with the glorious Tatiana Troyanos as Dido. And I treasure the 1993 recording with Lorraine Hunt Lieberson

as Dido, conducted by Nicholas McGegan. Readers interested in the libretto can find it at the very useful libretti site at Stanford University: http://opera.stanford.edu/iu/libretti/dido.html. (Recent scholarship has determined that the opera was composed earlier than the date given at the site — 1689; perhaps it was completed as early as 1684.)

12. Theocritus, "Syrinx," in *Greek Pastoral Poetry*, p. 197.
13. From Charles Michener, "The Soul Singer," in the *New Yorker*, January 5, 2004. Also available at http://www.newyorker.com/archive/2004/01/05/040105fa_fact.
14. These two passages are from Christoph Willibald Gluck's opera, *Orfeo ed Euridice* (the 1762 Vienna version), libretto by Ranieri de'Calzabigi. *"Che puro ciel"* is from Act Two and *"Che farò Euridice?"* is from Act Three.
15. Theocritus, "The Dioscuri," from *Greek Pastoral Poetry*, p. 121.

Pinus ponderosa: A Serious Waltz

1. Noted in Yves Cambefort, "Beetles as Religious Symbols," *Cultural Entomology, CE Digest*, Second issue (February 1994).
2. Material on Khepri and scarabs in general from Robert A. Armour, *Gods and Myths of Ancient Egypt*.
3. *El libro agregà de Serapiom, volgarizzamento di Frater Jacobus Philippus de Padua*. Ed. Gustav Ineichen. Venice and Rome: Instituto per la collaborazione culturale, 1966.
4. Theophrastus, *Enquiry into Plants*, p. 191.
5. Creber, G.I., "Tree Rings: A Natural Data-storage System", in *Biological Review* pp. 52, 354.
6. http://www.basketmakersco.org/.

Fagus sylvatica: Traces

1. Homer, *The Odyssey*, book 13, lines 404-06.
2. Glanville, *Schools of the Boundary*, pp. 96-97.
3. Owen, "Dulce et Decorum Est," in *Poems*, p. 14.

Arbutus menziesii: Makeup Secrets of the Byzantine Madonnas

1. Cennini, *The Craftsman's Handbook*, p. 45.
2. Ibid. p. 46.
3. Ibid. p. 84.
4. The discovery that Venetian artists had access to a wider range of ground glass than had previously been known is discussed by Barbara H. Berrie and Louisa C.

Matthew in "Material Innovation and Artistic Invention: New Materials and New Colours in Renaissance Venetian Paintings" in *Scientific Examination of Art*, pp. 12-28. A shorter account is given in by Alexandra Goho in "Venetian Grinds: The Secret Behind Italian Renaissance Painters' Brilliant Palettes," in *Science News*, March 12, 2005, pp. 168-169.
5. Kavanagh, *Collected Poems*, p. 130.

Populus tremuloides: Cariboo Wedding
1. Douglas, Meidinger, Pojar, eds. *Illustrated Flora of British Columbia*, Vol. 5.
2. Parish, Coupé, Lloyd, eds. *Plants of Southern Interior British Columbia*, p. 55.
3. The Stl'átl'lmx are the First Nations people of the middle Fraser Canyon or Lillooet area in British Columbia, members of the Interior Salish language family.
4. We had *Lorraine at Emmanuel* on the CD player. This is a wonderful recording celebrating Lorraine Hunt Lieberson and the late conductor Craig Smith with the Orchestra of Emmanuel Music in Boston, Massachusetts, released in 2008 by Emmanuel Music. The lines are from the aria "There in myrtle shades reclined" from Act 1, Scene 2 of Handel's *Hercules*.
5. Patrick Hannay died in 1629 after publishing several books, including *The Happy Husband* and *Two elegies on the late death of our Soveraigne Queene Anne, with Epitaphes*. This particular poem about the quaking aspen is cited in Maggie Campbell-Culver's *A Passion for Trees: The Legacy of John Evelyn*, p.135.

Arboretum: A Coda
1. The lines of poetry are from "Closing Down Kah Shakes Creek" by Charles Lillard, from his collection, *Shadow Weather*. Used by kind permission of Rhonda Batchelor Lillard.
2. I found salmonberry in bloom on February 17, 2010, but in 2011, I didn't see any until mid-March.
3. Stanley Kunitz, *The Collected Poems*, p. 217.
4. Cicero, *De Oratore*, p. 353.
5. John Evelyn, *Sylva*, p. 282.
6. Cicero, *Tusculan Disputations*, I, xxv, p. 65.
7. Ovid's line from *Ars Amatoria*, "*Silva domus ferat...cubilia fronds*" (Book 11, line 475), gracefully translated by my daughter, Angelica Pass. Source for the Latin: http://www.thelatinlibrary.com/ovid/ovid.artis2.shtml.

Bibliography

BOOKS

Angell, Tony. *Ravens, Crows, Magpies, and Jays*. Vancouver: Douglas & McIntyre, 1978.

Armour, Robert A. *Gods and Myths of Ancient Egypt*. Cairo: American University in Cairo Press, 1986.

Bachelard, Gaston. *The Poetics of Space*. Translated by Maria Jolas. Boston MA: The Beacon Press, 1994.

Boyd, Robert, ed. *Indians, Fire and the Land in the Pacific Northwest*. Corvallis OR: Oregon State University Press, 1999.

Budge, E. *The Book of the Dead: The hieroglyphic transcript and translation into English of the Papyrus of Ani*. Translated by A. Wallis. New York: Gramercy Books, 1995.

Campbell-Culver, Maggie. *A Passion for Trees: The Legacy of John Evelyn*. London: Eden Project Books, a division of Transworld Publishers, 2006.

Casselman, Bill. *Canadian Garden Words*. Toronto: Little, Brown and Company, Canada Ltd., 1997.

Cennini, Cennino D'Andrea. *Il Libro dell'Arte, or The Craftsman's Handbook*. Translated by D.V. Thompson. New Haven: Yale University Press, 1933.

Chaster, G.D., D.W. Ross, and W.H. Warren. *Trees of Greater Victoria: A Heritage*. Edited by J.W. Neill. Victoria BC: Heritage Tree Book Society, 1988.

Cicero, M.T. *On the Ideal Orator (De Oratore)*. Translated by E.W Sutton and H. Rackham. London and Cambridge: Harvard University Press, Loeb Classical Library, 1948.

___. *Tusculan Disputations*. Translated by J.E. King. Volume 18. London and Cambridge: Harvard University Press, Loeb Classical Library, 1927.

Clark, Lewis J. *Wild Flowers of the Pacific Northwest from Alaska to Northern California*. Edited by John G. Trelawny. Sidney BC: Gray's Publishing Limited, 1976.

Cocker, Mark. *Crow Country*. London: Jonathan Cape, 2007.

Cole, Douglas. *Captured Heritage: The Scramble for Northwest Coast Artifacts*. Norman OK: University of Oklahoma Press, 1995.

Cuppage, Judith. *Archaeological Survey of the Dingle Peninsula: A Description of the Field*

Antiquities of the Barony of Corca Dhuibhne from the Mesolithic Period to the 17th c. A.D. Dingle, Co. Kerry: Oichreacht Chorca Dhuibhne, 1986.

Dante. *The Inferno of Dante*. Translated by Robert Pinsky. New York: Farrar, Straus and Giroux, 1994.

da Vinci, Leonardo. *Leonardo on Painting: An Anthology of Writings by Leonardo Da Vinci with a selection of documents relating to his career as an artist*. Edited by Martin Kemp. Translated by Martin Kemp and Margaret Walker. New Haven and London: Yale University Press, 1989.

Deur, Douglas and Nancy J. Turner, eds. *Keeping in Living: Traditions of Plant Use and Cultivation on the Northwest Coast of British Columbia*. Seattle, Toronto, Vancouver: University of Washington Press, UBC Press, 2005.

Dihl, Albrecht. *A History of Greek Literature from Homer to the Hellenistic Period*. Translated by Clare Krojzl. London and New York: Routledge, 1994.

Dioscorides, Pedanius. *The Greek Herbal of Dioscorides, illustrated by a Byzantine, A.D. 512*. Translated by John Goodyer. New York and London: Hafner Publishing, 1959.

Douglas, G.W., D. Meidinger, and J. Pojar, ed. *Illustrated Flora of British Columbia*. Dicotyledons (Salicaceae through Zygophyllaceae) and Pteridophytes. Volume 5. Victoria: BC Minist. Environ., Lands and Parks, and BC Minist. For., 2000.

Duff, Wilson. *Thunderbird Park*. Revised edition. Victoria: BC Government Travel Bureau, 1963.

Durrell, Lawrence. *Reflections on a Marine Venus*. London: Faber and Faber Co., 1960.

Elytis, Odysseas. *Eros, Eros, Eros: Selected and Last Poems*. Translated by Olga Broumas. Port Townsend WA: Copper Canyon Press, 1998.

Evelyn, John. *Sylva: A Discourse of Forest Trees & the Propagation of Timber, Volumes 1 & 2*. New York: Doubleday and Co., 1908.

Fletcher, Neville H. *Acoustic Systems in Biology*. New York: Oxford University Press, 1992.

Franklin, Douglas and Martin Segger. *Exploring Victoria's Architecture*. Victoria BC: Sono Nis Press, 1996.

Gibbon, Edward. *The History of the Decline and Fall of the Roman Empire*. Harmondsworth: Penguin, 2001.

Glanville, Alice. *Schools of the Boundary: 1891-1991*. Merritt BC: Sonotek Publishing Ltd. 1991.

Glazebrook, George Parkin de Twenebroker, ed. *The Hargrave Correspondence, 1821-43*. Toronto: Champlain Society, 1938.

Green, Miranda, ed. *The Celtic World*. London: Routledge, 1995.

Harbison, Peter. *Ancient Irish Monuments*. Dublin: Gill and Macmillan, 1997.

Hare, Jan and Jean Barman, *Good Intentions Gone Awry: Emma Crosby and the Methodist Mission on the Northwest Coast*. Vancouver: UBC Press, 2006. All rights reserved by the publisher.

Hayden, Brian, ed. *A Complex Culture of the British Columbia Plateau: Traditional* Stl'átl'imx *Resource Use*. Vancouver: UBC Press, 1992.

Heinrich, Bernd. *Mind of the Raven: Investigations and Adventures with Wolf-Birds*. New York: Harper Collins, 2002.

___. *Ravens in Winter*. New York: Vintage Books, 1989.

Herodotus. *The Histories (The Landmark Herodotus)*. Edited by Robert B. Strassler. Translated by Andrea L. Purvis. New York: Pantheon Books, 2007.

___. *The Histories*. Translated by George Rawlinson. London: J. M. Dent & Sons, 1992.

Hesiod. *Homeric Hymns and Homerica*. Translated by Hugh G. Evelyn-White. Cambridge MA: Harvard University Press, 1914.

Hopkins, Gerard Manley. *Poems*. Edited by Robert Seymour Bridges. Oxford: Oxford University Press, 1948.

Homer. *The Odyssey*. Translated by Robert Fitzgerald. New York: Doubleday, 1963.

House, Syd and Ann Lindsay. *The Tree Collector: The Life and Explorations of David Douglas*. London: The Aurum Press, 2005.

Howatson, M.C., ed. *The Oxford Companion to Classical Literature, second edition*. New York, Oxford: Oxford University Press, 1989.

Kavanagh, Patrick. *Collected Poems*. London: Allen Lane, 2004.

Kunitz, Stanley. *The Collected Poems*. New York: W.W. Norton, 2000.

Lillard, Charles. *Shadow Weather*. Victoria: Sono Nis Press, 1996.

MacKillop, James. *A Dictionary of Celtic Mythology*. Oxford: Oxford University Press, 2004.

MacKinnon, Andy and Jim Pojar, ed. *Plants of Coastal British Columbia including Washington, Oregon & Alaska (revised)*. Vancouver, Edmonton, Auburn WA: Lone Pine Publishing, 2004.

McManus, Damian. *A Guide to Ogam*. Maynooth, Eire: An Sagart, 1991.

McFarland, Jeannie. *Pine Needle Raffia Basketry*. Provo UT: Press Publishing, Ltd., 1978.

Mabey, Richard. *Beechcombings: The Narratives of Trees*. London: Chatto and Windus, 2007.

Marzluff, John M. *In the Company of Crows and Ravens*. Illustrated by Tony Angell, with a foreword by Paul R. Erlich. New Haven: Yale University Press, 2005.

Miyazaki, Masajiro. *My Sixty Years in Canada*. Lillooet BC: Privately printed, 1973.

Miclea, Ion. *Sweet Bucovina*. Romania: Editura Sport-Turism, 1977.

Miller, Henry. *The Colossus of Maroussi*. New York: New Directions, 1958.

Miller, Stephen G. *Arete: Greek Sports from Ancient Sources*. Berkeley & Los Angeles: University of California Press, 2004.

Milne, Lorus and Margery Milne. *National Audubon Society Field Guide to North American Insects and Spiders*. New York: Alfred A. Knopf, Inc., 2000.

Nuytten, Phil. *The Totem Carvers: Charlie James, Ellen Neel, and Mungo Martin*. Vancouver: Panorama Publications, Ltd., 1982.

O'Kelly, Michael. *Early Ireland: An Introduction to Irish Prehistory*. Cambridge: Cambridge University Press, 1989.
Orchard, Andy. *Cassell's Dictionary of Norse Myth and Legend*. London: Cassell & Co., 2003.
Owen, Wilfred. *Poems*. Whitefish MT: Kessinger Publishing, 2004.
Pakenham, Thomas. *Meetings with Remarkable Trees*. London: Cassells Paperback, 2001.
Parish, Roberta, Ray Coupé, and Dennis Lloyd. *Plants of Southern Interior British Columbia and Inland Northwest*. Vancouver, Edmonton, Renton WA: Lone Pine Publishing, 1996.
Paton, John Glenn, ed. *26 Italian Songs and Arias: An Authoritative Edition Based On Authentic Sources*. Alfred Publishing, Ltd., 1991.
Pavord, Anna. *The Naming of Names: The Search for Order in the World of Plants*. New York: Bloomsbury Publishing, 2005.
Payne, Barbara A., ed. dir., *Field Guide to the Birds of North America*. Washington DC: National Geographic Society, 2000.
Pethick, Derek. *Summer of Promise, Victoria 1864-1914*. Victoria: Sono Nis Press, 1980.
Poulton, Diana. *John Dowland: His Life and Works*. Berkeley & Los Angeles: University of California Press, 1982.
Pyne, Stephen J. *Fire: a Brief History*. Seattle and London: University of Washington Press, 2001.
Savage, Candace. *Bird Brains: The Intelligence of Crows, Ravens, Magpies and Jays*. Vancouver BC: Greystone Books, 1995.
Seferis, George. *Collected Poems*. Translated by Edmund Keeley and Philip Sherrard. London: Anvil Press Poetry Ltd., 1986.
Snyder, Gary. *The Practice of the Wild*. Berkeley: North Point Press, 1990.
Steinberg, S.H. *Five Hundred Years of Printing*. Harmondsworth: Penguin, 1961.
Swanton, John R. *The Indians of the Southeastern United States, Bureau of American Ethnology Bulletin 137*. Washington, DC: Smithsonian Institution, 1979.
Theocritus. "Syrinx." In *Greek Pastoral Poetry*. Translated by Anthony Holden. Harmondsworth: Penguin, 1974.
Theophrastus. *Enquiry into Plants*. Translated by Arthur Hort. Cambridge MA: Loeb Classical Library, Harvard University Press, 1919.
Virgil. *Virgil's Georgics: A Modern English Verse Translation*. Translated by Smith Palmer Bovie. Chicago: University of Chicago Press, 1966.
Ward, Robin. *Echoes of Empire: Victoria & its Remarkable Buildings*. Madeira Park, BC: Harbour Publishing, 1996.
Ward-Harris, J. *More Than Meets The Eye: The Life and Lore of Western Wildflowers*. Toronto: Oxford University Press, 1983.
Whitman, Walt. *Complete Poetry and Collected Prose*. New York: Penguin Putnam, 1982.
Wolff, Tobias. *This Boy's Life*. New York: Grove/Atlantic, 2000.

Wyse-Jackson, Peter. *Irish Trees and Shrubs*. Belfast: Appletree Press, 1994.
Yates, Frances A. *The Art of Memory*. London: Ark Editions, 1984.

ARTICLES

Berrie, Barbara H. and Louisa C. Matthew. "Material Innovation and Artistic Invention: New Materials and New Colours in Renaissance Venetian Paintings." In *Scientific Examination of Art: Modern Techniques in Conservation and Analysis*, from Arthur M. Sackler Colloquim held at the National Academy of Sciences building. Washington DC: National Academies Press, 2005: pp. 12-28.

Bookidis, Nancy. "The Sanctuary of Demeter and Kore on Acrocorinth, Preliminary Report." *Hesperia* 38, no.3 (July-September, 1969): pp. 297-310.

Cambefort, Yves. "Beetles as Religious Symbols." *Cultural Entomology, CE Digest*, Second Issue (February 1994). (Accessed Fall 2008) <http://www.insects.org/ced1/beetles_rel_sym.html>.

Coghlan, Andy. "Sensitive Flower." *New Scientist* (September 26, 1998): pp. 23-24.

Creber, G.T. "Tree Rings: A Natural Data-storage System." *Biological Review* 52 (1977): p. 354.

Esson, Heather, Aaron Heiss, Chris Sears, and Nyssa Temmel. "Comparison of Two Garry Oak Sites Undergoing Restoration on Southeastern Vancouver Island: A Preliminary Study." *Davidsonia* 17, no. 3 (March 2007): pp. 87-97.

Flematti, Gavin, R., Emilio L. Ghisalberti, Kingsley W. Dixon, and Robert D. Trengove. "A Compound from Smoke That Promotes Seed Germination." *Science* vol. 305, no. 5686 (August 2004): p. 977.

Goho, Alexandra. "Venetian Grinds: The Secret Behind Italian Renaissance Painters' Brilliant Palettes." *Science News* 167 (March 12, 2005): pp. 168-169.

Gruber, Jacob W. "Ethnographic Salvage and the Shaping of Anthropology." *American Anthropologist* 72, no. 6 (1970): pp. 1289-99.

Jacknis, Ira. "Authenticity and the Mungo Martin House, Victoria, B.C.: Visual and Verbal Sources." *Arctic Anthropology* 27, no. 2 (1990): pp. 1-12.

Keeley, J.E. and C.J. Fotheringham. "Role of fire in regeneration from seed." *Seeds: The Ecology of Regeneration in Plant Communities*. 2nd Edition. Edited by M. Fenner. Oxon UK: CAB International, 2000, pp. 311-330.

Lea, T. "Historical Garry Oak Ecosystems of Vancouver Island, British Columbia, pre-European Contact to the Present." *Davidsonia* 17, no. 2 (2006): pp. 34-50.

Michener, Charles "The Soul Singer." *The New Yorker*. January 5, 2004. Also available online (Accessed May 2012) <http://www.newyorker.com/archive/2004/01/05/040105fa_fact?currentPage=all>.

Moffett, Mark W. "What's 'Up'? A Critical Look at the Basic Terms of Canopy Biology." *Biotropica* 32, no. 4a (December 2000): pp. 569-596.

Playfair, A.S. "Hippocratic Plane." *Journal of the Royal College of General Practitioners* 107 (June 21, 1971): pp. 367-368.

Schulman, Edmund. "Tree-Rings and History in the Western United States." *Economic Botany* 8, no.3, (1954): pp. 234-250.

Slane, Kathleen Warner. "The End of the Sanctuary of Demeter and Kore on Acrocorinth." *Hesperia* 77, no.3 (July–September 2008): pp. 465-496.

Stubbings, Frank H. "Xerxes and the Plane-Tree." *Greece & Rome* 15, no. 44 (May 1946): pp. 63-67.

Titze, Ingo R. "The Human Instrument." *Scientific American* (January 2008): pp. 94-101.

WEB SITES

Blane, Douglas. "Language of trees branches into poetry". (Accessed February 2007) http://www.tes.co.uk/article.aspx?storycode=2115109

Botanical Electronic News, #226, July 2, 1999. (Accessed February 2008) http://www.ou.edu/cas/botany-micro/ben/ben226.html

British Colonist Online Edition: 1858-1910. (Accessed November 2008) http://britishcolonist.ca/

British Columbia Protestant Orphans' Home. (Accessed March 2007) http://web.uvic.ca/vv/student/orphans/home.html

Bukovina, Heaven On Earth. (Accessed October 2009) http://heaven-bukovina.blogspot.com/2009/02/orthodox-psalms-at-putna-monastery.html

Cornet, Bruce, Upper Cretaceous Facies, Fossil Plants, Amber, Insects and Dinosaur Bones, Sayreville, New Jersey. (Accessed February 2010) http://www.sunstar-solutions.com/sunstar/Sayreville/Kfacies.htm

Economic Botany Collection at Royal Botanic Gardens, Kew. (Accessed January 2009) http://www.kew.org/collections/ecbot/collections/region/canadian-aboriginal-artefacts/index.html)

Garry oak ecosystems recovery team. (Accessed February 2007) http://www.goert.ca

Guardian News Online: http://www.guardian.co.uk/

Klinkenberg, Brian, ed. E-Flora BC: Electronic Atlas of the Plants of British Columbia. Lab for Advanced Spatial Analysis, Department of Geography, University of British Columbia, Vancouver. (Accessed August 2010) http://www.geog.ubc.ca/biodiversity/eflora/

Libretto List. (Accessed March 2011) http://opera.stanford.edu/iu/libmlist.html

Metcalf, Alex. "Tree Listening". (Accessed September 2008) http://www.alexmetcalf.co.uk

Michener, Charles, "The Soul Singer". (Accessed August 2006) http://www.newyorker.com/archive/2004/01/05/040105fa_fact

Ovid, Ars Amatoria. (Accessed February 2011) http://www.thelatinlibrary.com/ovid/ovid.artis2.shtml

Ringuette, Janis. Beacon Hill Park History, 1842-2009. (Accessed January 2009) www.beaconhillparkhistory.org

Worshipful Company of Basketmakers. (Accessed November 2008) http://www.basketmakersco.org/

Theresa Kishkan was born in Victoria, BC, and has lived on both coasts of Canada as well as in Greece, England, and Ireland. She makes her home on the Sechelt Peninsula with her husband, John Pass, where they built their house and raised their three children. Together, they operate High Ground Press, which prints broadsheets and chapbooks on a nineteenth-century platen press.

Kishkan is the author of eleven books of poetry and prose, including two collections of essays, *Red Laredo Boots* and *Phantom Limb*; three novels, *Sisters of Grass*, *A Man in a Distant Field*, and *The Age of Water Lilies*; and a novella, *Inishbream*. Her writing has appeared in numerous magazines and been nominated for several prestigious awards, including the Pushcart Prize, the Ethel Wilson Prize for Fiction, the Hubert Evans Prize for Non-Fiction, and the ReLit Award. *Phantom Limb* won the inaugural Readers' Choice Award given by the Canadian Creative Non-Fiction Collective. An essay from *Mnemonic* won the 2010 Edna Staebler Personal Essay Prize.